Coherent Optics for Access Networks

Coherent Optics for
Access Networks

Edited by
Zhensheng Jia and Luis Alberto Campos

CRC Press
Taylor & Francis Group
Boca Raton London New York

CRC Press is an imprint of the
Taylor & Francis Group, an **informa** business

CRC Press
Taylor & Francis Group
6000 Broken Sound Parkway NW, Suite 300
Boca Raton, FL 33487-2742

First issued in paperback 2023

© 2020 by Taylor & Francis Group, LLC
CRC Press is an imprint of Taylor & Francis Group, an Informa business

No claim to original U.S. Government works

ISBN-13: 978-0-367-24576-4 (hbk)
ISBN-13: 978-1-03-265456-0 (pbk)
ISBN-13: 978-0-429-28410-6 (ebk)

DOI: 10.1201/9780429284106

Visit the Taylor & Francis Web site at
http://www.taylorandfrancis.com

and the CRC Press Web site at
http://www.crcpress.com

Zhensheng Jia dedicates this book to his wife Rui and his two daughters Surie and Annie.

Luis Alberto Campos dedicates this book to his wife Sara and his two daughters Nicole and Jacqueline.

Without their love and support this book would not be possible.

Contents

Foreword

The transport of optical fiber has addressed the requirements of system capacity and transmission data rates since the inception of the Internet. The development of both low-loss fiber and advanced optical transmitters and receivers has made high-capacity, long-distance networks more practical and deployable. Advancements in fiber manufacturing have made great strides toward reducing loss in fiber, controlling and lowering dispersion in addition to extending the range of linear operations through a larger effective fiber core area with better power-handling capabilities.

Around 2008, optical receiver implementations began leveraging signal-processing capabilities to make coherent reception practical. Today, the telecommunication industry has predominantly used coherent optics technology in long-haul and is, more recently, taking hold in metro applications. The authors of this book had the foresight to explore the challenges of launching coherent optics technology in the next frontier of optical access networking environment. The major challenges here are how to redesign and optimize the performance of the system based on the shorter link distances of point-to-point and point-to-multipoints (P2MPs) of access system, one that leverages its significant link margin for greater flexibility in relaxed performance specifications of both the components and systems.

The authors' deep-rooted association with CableLabs and the telecom industry affords them with a unique perspective derived from a close partnership with service providers and component manufacturers. This provides them with valuable insights into the evolution of network architectures, broad knowledge about types of services, availability of resources for deployment scenarios and demand for capacity. The long-term track records of the authors also place them at the forefront in developing leading-edge optical and broadband access technologies and network architectures.

The topics covered in this book are very timely in which the authors are helping the optical data transport industry to prepare for the critical transition toward coherent optics. The scale of economic impact in the access networks could become much greater than that of long-haul and metro optical transport networks.

This book does a superior job at differentiating the challenges confronted by access infrastructure builders from the ones that are found in the long-haul, metro and even data center interconnect networks. In access networks, minimizing the implementation costs is a key requirement. The authors embrace an end-to-end economic-analysis approach that provides insight for the total cost of ownership of such a forward-looking network technology.

As discussed in this book, the deployment of coherent optics in optical transport technology for long-haul, metro and access networks are emerging just like the introduction of dense wavelength division multiplexing (DWDM) technology three decades ago. It is important to analyze and discuss the network operation based on the fundamental requirement to carry all types of services. This emerging multi-service network ideally would meet the performance requirements to support all high-speed, low-latency, high-reliability services at low cost.

The evolution of mobile data networks toward 5G will require a very high capacity and dense optical distribution network in order to deliver a wide diversity of use cases. The overall cost per bit of this future backhaul/midhaul/fronthaul network needs to be minimized while achieving these goals. The optical-access technologies discussed in this book provide a promising platform to connect the converged optical and wireless networks of the future.

Traditional intensity-modulation and direct-detection (IM-DD) optical P2MP networks are reaching the limits of speed and distance that can be achieved. This book provides a firsthand view of what the passive optical network (PON) network of the future will look like after the coherent PON networks are introduced. Coherent PON technology is a promising solution in significantly extending reach, capacity and split ratio.

The exponential growth in capacity requires a technology that can be flexibly expanded as hot spots of demand arise. These future access networks have to be scalable with a growth in capabilities anticipating this exponential growth. The introduction of coherent optics in the access environment can meet the capacity challenges originating from the limited lifespan of legacy technologies.

The authors' view of access networks provides valuable insights into solving future network capacity challenges with coherent optics as well as enabling other leading-edge technologies and advanced network architectures.

Dr. Gee-Kung Chang
Georgia Research Alliance and Byers Eminent Scholar Chair
Professor in Optical Networking
School of Electrical and Computer Engineering
Georgia Institute of Technology
Atlanta, Georgia
USA

Preface

The year 2020 is digital coherent optical technology's 10th anniversary of being officially deployed in long-haul applications. Today, coherent optical technology focuses mainly on two development areas: high-end programmable coherent transceivers capable of processing 100–800G data rates per single wavelength and power-consumption reduction to meet the size and cost requirements of access applications.

The former increases the spectral efficiency and signal band rate to maximize single wavelength capacity. It leverages more complex modulation formats and powerful forward-error-correction transmission efficiency and cost per bit is further improved. In the latter, advancements in complementary metal–oxide–semiconductor (CMOS) node process, reduction in design complexity and the price drop in opto-electronic components have moved coherent solutions from long-haul and metro to access networks. These implementations support high bandwidth demand through lowering the power, size and cost of coherent technology for use in pluggable applications. The market for coherent optical links of lengths between 10 and 120 km is emerging for many application scenarios, such as telco metro-access router-to-router interconnects, point-to-point (P2P) data-center interconnect, and mobile- and cable-aggregation applications.

Access networks have been steadily evolving in their capabilities, architectures and the type of services they carry. Telephony access networks have consisted not only sizeable fiber to the home deployments but also greater numbers of remote digital subscriber line access multiplexers (DSLAMs) and higher capacity very high speed digital subscriber line (VDSL) and G.fast technologies. Cellular networks are now introducing 5G technologies which not only promise to deliver higher speeds but also demand deeper, denser fiber-distribution networks. Cable has been deploying fiber deeper and migrating toward distributed access architectures that bring fiber much closer to subscribers. All networks have in common a transition toward much greater capacity per subscriber, fiber that is in closer proximity to subscribers and the delivery of all types of services. Coherent optics is a very suitable technology to address the long-term evolution of access networks.

Coherent detection for access networks enables the superior receiver sensitivity that allows for an extended power budget. The high spectral efficiency in coherent optics enables the maximum use of legacy fiber infrastructure and future-proof network upgrades. In long-distance transmission, the use of discrete performance-optimized optical components and best in class proprietary digital signal processing algorithms was critically important to gain a competitive advantage in coherent optical deployments. However, it is over-engineered, too expensive and too power-hungry when the coherent optics extends its application into the access. The access network is a totally different environment compared to long-haul and metro. The reduced link-budget requirements enable a much more simplified coherent transceiver design with low-power consumption. Access network may need a hardened solution for remote site location, in which the temperature is not controlled. It is highly advantageous that coherent transceivers from different system vendors be interoperable at the optical transport layer to simplify the deployment of multivendor networks. The standardization of coherent interconnects has been making significant progress in the last 2 years. Standardization will eventually lead to improved interoperability and predictable performance, which are the keys to providing a low-cost solution using coherent optics.

In this book, the recent developments in the field of coherent optics for access network applications that will support P2P aggregation use cases and point-to-multipoints (P2MP) fiber to the user passive optical network are examined. The book presents the trends of optical industry as well as the conventional IM-DD systems and newly developed advanced direct-detection architectures leveraging 4-level pulse amplitude modulation format, Stokes receivers and Kramers–Krönig receivers. This book focuses on coherent optics technology and how we adapt it to the access environment in ways that address major cost challenges, such as simplified transceiver design and photonic integration. This book analyzes both P2P and P2MP use cases and provides the economical modeling for aggregation case with the comparison of 10G IM-DD DWDM solution. Implementation requirements, unique to the access environment, are also provided when introducing coherent optics into access scenarios, including bidirectional single fiber connection, coexistence with existing services and security challenges. Progress on recent-specification development activities is reviewed for many industry organizations including Optical Internetworking Forum, CableLabs, OpenROADM, International Telecommunication Union-T, and Institute of Electrical and Electronics Engineers which focus on short-distance coherent optics interoperability. This book is meant to be an introduction for any reader interested in coherent optical technology and its expanded use case in access networks and to the readers interested in optical industry trends and standardization development on the advancement of coherent optics technology.

Acknowledgments

We are extremely grateful to our colleagues at CableLabs, CableLabs' P2P Coherent Optics working group members, and our operator members, who have motivated, inspired and guided us to work in this new field of coherent optics for access networks.

We also wish to express our thanks to Prof. Gee-Kung Chang for his insights and guidance and to the publishing team for their effort and patience in order to bring this project to fruition.

Also, special thanks to all the authors for contributing chapters for this book.

Zhensheng Jia
Distinguished Technologist
Cable Television Laboratories, Inc.

Luis Alberto Campos
Fellow
Cable Television Laboratories, Inc.

Editors

Zhensheng Jia, PhD, is a Distinguished Technologist and Technical Lead of multiple innovation and R&D projects for optical access networks. He has published more than 160 peer-reviewed papers and has more than 20 granted patents. He earned his PhD in Electrical and Computer Engineering at the Georgia Institute of Technology.

Luis Alberto Campos, PhD, is First Fellow in CableLabs' history and has 33 granted patents and numerous publications. He earned his PhD in Electrical Engineering from Northeastern University.

Contributors

Lin Cheng, PhD
Office of the CTO
Cable Television Laboratories, Inc.
Louisville, Colorado

Mustafa Sezer Erkilinc, PhD
Photonics Networks and Systems
    Department
Fraunhofer Institute for
    Telecommunications, HHI
Berlin, Germany

Curtis Knittle, PhD
Research and Development Department
Cable Television Laboratories, Inc.
Louisville, Colorado

Domanic Lavery, PhD
Department of Electronic & Electrical
    Engineering
University College London
London, United Kingdom

Chris Stengrim
Strategy Department
Cable Television Laboratories, Inc.
Louisville, Colorado

Jing Wang, PhD
Office of the CTO
Cable Television Laboratories, Inc.
Louisville, Colorado

Mu Xu, PhD
Office of the CTO
Cable Television Laboratories, Inc.
Louisville, Colorado

Haipeng Zhang, PhD
Research and Development Department
Cable Television Laboratories, Inc.
Louisville, Colorado

Junwen Zhang, PhD
Office of the CTO
Cable Television Laboratories, Inc.
Louisville, Colorado

1 Access Networks Evolution

Luis Alberto Campos and Zhensheng Jia

CONTENTS

1.1 INTRODUCTION

There has been an amazing growth in data networks since the early days of the Internet. In the beginning, the Internet consisted of a network of connected servers at educational and industrial locations that users would access by being directly connected to the server on site or indirectly through remote terminals with modems connected to the main computers through telephone lines. These remote telephony access links represent the very first access networks. These networks have grown at exponential rates to the point that today they cover more than 3 billion users worldwide connected to the Internet across fixed and wireless networks [1]. The predominant way subscribers access the Internet is through a service provider's access network.

Access networks can either be connected through wires or be wireless. Wired access networks connect subscribers via twisted pair of wires, coaxial transmission lines, through fiber and even through power lines. Wireless access networks provide Internet access through the cellular mobile infrastructure, wireless access points using WiFi, fixed networks that rely on static point-to-point wireless connections, and satellite connectivity.

The capability of these networks has also been growing to keep up with the exponential growth of demand for capacity. By extrapolating Nielsen's law of Internet bandwidth, broadband service rates are expected to reach 10 Gbps by 2023 and 100 Gbps by 2029 [2].

Access networks have been originally designed for the delivery of specific services but have been adapted for Internet connectivity.

In all cases, providers have optimized cost as they have leveraged much of their existing infrastructure to provide Internet connectivity. Evolution of networks through fiber optic links has played a greater role in meeting the increased demand

1

for capacity. Optical transport has been a tool used by telephony, cable TV and mobile service providers. These networks have introduced more and more fiber as the demand is keep increasing. Service providers have not limited themselves to provide connectivity to the masses, but they have also provided connectivity to enterprises, building metropolitan and regional aggregation networks.

This chapter analyzes different types of optical transport and access network architectures. The optical transport technologies that have been dominant in the access environment such as analog optics and baseband digital optics leveraging intensity-modulation and direct-detection (IM-DD) are also discussed. The characteristics and limitations of these technologies are important to understand the drivers for introducing coherent optics in the access network as a long-term technology strategy. Current use of point-to-point (P2P) and point-to-multipoint (P2MP) technologies in the access is also discussed.

1.2 TRAFFIC GROWTH

Video-intensive technologies require the most bandwidth, and immersive applications leveraging Virtual Reality/Augmented Reality (VR/AR) are the most demanding out of all video applications. Current VR applications are little more than 360° video/panoramas. A low-quality 360° video requires at least a 30 Mbps connection, HD quality streams easily surpass 100 Mbps, and retina quality (4k+) streams approach Gbps territory. Looking one step ahead in the evolution of immersive videos, holographic displays or light-field displays are being prototyped. As the name indicates, light-field displays reproduce the original light fields through a display, generating the original electromagnetic waves that would have been generated from actual objects from a display window. This allows the viewer to move and change viewing angles and perceive the natural changes of the image, such as seeing objects behind the foreground appear with viewing angle changes. Light-field displays provide true 3D and immersive visual representation without headsets. It is estimated that a commercial size display would require a capacity around 1.5 Gbps [3].

However, there are still many things holding back its use beyond showrooms and proof of concepts, the most glaring problem being the network's capacity.

These days, it seems that just about everything is getting smarter, from thermostats to refrigerators, and becoming ever more connected. Each of these devices – physical objects with data sensing, analyzing and recording functions plus the ability to communicate remotely – collectively forms the "Internet of Things" (IoT). Clearly, expansion in the use of smart devices is an unstoppable force, but one thing could hamper this growth – inadequate bandwidth. While most of the devices that comprise the IoT communicate wirelessly all the world over, all the data they send must be transmitted over a physical wireline network between wireless access points. High-throughput, low-latency and high-reliability networks will be needed for applications such as video analytics in public safety and to support self-driving cars.

In the upcoming 5G era, the impacts of massive MIMO (Multiple-Input Multiple-Output), Carrier Aggregation, Multi-band support and radio cell densification necessitate bandwidth requirements, while the impact of coexistence between macro-, micro-, pico- and small cells in a centralized/virtualized processing environment calls for flexibility requirements. Fiber and optical access technologies are expected to play more and more important roles in the fronthaul and backhaul services to meet the aggressive performance goals of 5G.

Access networks have been evolving to meet the ever-increasing demand for capacity. Three dominant approaches have been used to address the increasing demand for capacity: first, increasing optical transport efficiency with greater capacity per bandwidth; second, segmenting network into smaller serving areas and smaller wireless cells to dedicate the same target capacity to fewer end devices and to be closer to subscribers for higher transport performance; and third, adding more spectrum. The fourth, least desired and more costly approach entails deploying more fiber.

1.3   TELCO, MOBILE AND CABLE ACCESS NETWORKS

An access network is the one that a network operator uses to provide connectivity to a large number of subscribers. Subscriber connectivity is provided through an aggregation and distribution center. Telephone and cellular aggregation and distribution centers are called central offices, while in the case of cable, they are often referred to as hubs.

While telephone, cellular and cable networks were, in the past, differentiated from each other in topology, architecture and the services they carry, in recent years, an evolution from different starting points, converging toward the delivery of IP services, has resulted in these networks resembling each other as they continue to evolve.

Earlier, cable networks focused on delivery of video services through end-to-end coaxial transport, but later on, they evolved toward a bidirectional hybrid fiber coaxial (HFC) network to carry data, voice and video services.

Telephone networks initially focused on telephony services through point-to-point twisted pair connectivity from central office to subscribers. Analog voice transport evolved to 56/64 kbps digital voice, and this evolved again into digital subscriber line (DSL) services, leveraging twisted pair infrastructure with digital signal processing to achieve high data-carrying capacity.

Cellular networks consisted of macro-cells to deliver wireless telephony services, but by augmenting the spectrum they use and increasing the number of cells, cellular networks evolved toward the transport of high-bandwidth data in addition to voice services.

All of the above networks, in their quest to meet ever-increasing capacity requirements of subscribers, have deployed more fiber in their networks.

Many of these service providers have introduced fiber-to-the-home (FTTH) networks to address capacity issues in high-demand areas. These fiber networks are also known as passive optical networks or PONs. The dominant digital 10G PONs standards are XG-PON/XGS-PON [4,5] and 10G- Ethernet passive optical network (10G-EPON) [6]. XG-PON supports 10 Gbps downlink (DL) and 2.5 Gbps uplink (UL), XGS-PON provides symmetrical 10 Gbps speed, while 10G-EPON supports 10 Gbps DL and 1.25 Gbps UL. In these PON systems, the network transitions from one optical line termination (OLT) unit to 32 or 64 optical end-points or optical network units (ONUs) typically traverse a distance of no more than 20 km. Today, PON technology uses optical IM-DD schemes. The sensitivity of IM-DD limits the reach of PON to 20 km.

1.4   DATA CENTER INTERCONNECTIVITY NETWORKS

Data center networks are very different. Even though many data center networks may be deployed in metropolitan area settings and may share similar distances as those found in access networks, they should not be considered access networks. Data centers and their networks have different growth characteristics, both in the number of links that interconnect data centers and in the capacities of those links. Data center topologies are also quite different from access network topologies. While access networks for the most part are implemented to establish connectivity from one to a very large number of end-points, data center networks are implemented establishing full mesh interconnectivity between fewer data center end-points. The capacity between any two data center end-points is many times massive, requiring a large number of parallel links to establish connectivity between the different data center end-points.

1.5   ACCESS NETWORKS CHARACTERISTICS

Service selection, transmission medium and transport technologies unique to each type of access network have implications on the network size and capabilities, on how these access networks have been deployed and how they have evolved. Some details of how the different access technologies correspond to and differ from each other in these areas, including the fiber infrastructure associated with each type of access technology, are discussed in the following.

Telcos have fiber links that connect remote terminal digital subscriber line access multiplexers (DSLAMs) and remote G.fast units, which are capable of much greater speed than central office (CO)–based DSLAMs when in closer proximity to subscribers. Cellular companies have evolved from macro-cells to small cells to increase capacity/cell, while cable companies have evolved to distributed access architectures to increase capacity per node and transport efficiency. The next-generation transport architectures of both cellular and cable networks incorporate function splits that impact not only optical network topology but also optical transport resources.

Today, optical backhaul capacity to these smaller serving areas is provided through PON technology or point-to-point gigabit ethernet. These fiber links typically carry no more than 10 Gbps to the aggregation center such as a central office or a hub location. Newer PON standards are aimed to extend the capacity up to 50 Gbps [7].

Remote DSLAMs should be able to handle more than 500 users, while fiber-to-the-distribution-point (FTTdp) and G.Fast nodes would handle fewer users consistent with the shorter loop reach.

In cable, subscribers within a serving group are, in general, served from a coaxial network that extends from a fiber node. In cable, the serving area size is not dependent on the distance between the fiber node and subscribers, but the serving area size is, for the most part, decided on aggregate capacity that is required to provide a mix of services for the total number of subscribers within the serving area. By fully leveraging all the RF spectrum currently available in cable, an aggregate downstream capacity of 10 Gbps can be achieved. Today in the United States, cable is capable to provide 1 Gbps service to 80% of their subscribers [8].

In order to meet a traffic demand of 10 Gbps per end-user, prevalent IM-DD optical transport technologies, such as 10G Ethernet or 10G PON, are approaching their capacity limit for even one end device let alone supporting an aggregate number of high-capacity end devices. The optical transport network has to evolve and be capable to handle the access network needs of greater backhaul capacity.

1.6 FIBER DEPLOYMENT IN THE ACCESS

Plain old telephone service (POTS) is provided from central offices through twisted pair links typically traversing no more than 6 km (20 kft) to reach end subscribers. Most central offices have good fiber connectivity, while additional fiber has been installed to enhance broadband services through broadband nodes closer to subscribers such as remote DSLAMs, very high speed DSL (VDSL) nodes and G.fast FTTdp nodes. Vectored VDSL nodes, as close as 1,000 ft from subscribers, deliver speeds beyond 100 Mbps, while G.fast nodes within 300 ft from subscribers deliver speeds beyond 500 Mbps. In the United States, for example, the Telco fiber network reaches more than 25,000 central offices. In addition, there are many remote broadband nodes connected through fiberoptic links that extend from the central office [9]. This fiber network including the FTTH deployments that reach customers directly with fiber and the fiber that has been deployed for enterprise connectivity is indicative of the extensive Telco's fiber coverage.

So far, cellular's fiber network has focused on connecting macro- and small cells extending from the macro-cells in areas of high traffic demand. The cellular industry, however, is gearing toward a very dense deployment of cells as wireless networks evolve to 5G and a very deep deployment of fiber supporting 5G backhaul, midhaul and fronthaul. In 5G, hierarchical optical distribution networks are expected to be deployed to provide interconnectivity among centralized units and distributed units [10,11]. In the United States, there are more than 320,000 cell sites [12], and the average of number of cell sites for the four largest cellular service providers is greater than 60,000. A large number of small cells are targeted for deployment to supplement the capacity needed within a macro–cell. The demand in fiber connectivity to meet the capacity per cell site and the growth in the number of cell sites is driving a significant deployment of fiber.

In cable, the HFC network from the hub to the subscriber is considered the access network. Ninety percent of fiber links from hub to node are less than 40 km. In the United States, for example, there are more than 5,000 hubs from which broadband services are available to more than 90% of US households [7]. Metropolitan and regional fiber networks connect these hubs that are divided in fiber node serving areas, each typically covering 400–500 households. Today, the average distance between cable optical terminals or fiber nodes and subscribers is less than 1 km. This last segment of coaxial network still leverages a few amplifiers in cascade to reach subscribers. Cable migration of significant portions of its HFC network to fiber deeper architectures called N+2 and N+0 places around 10–30 optical terminal points every km^2 in suburban areas. As a result, in these fiber deeper network architectures, any subscriber is within a range of 100–200 m from an optical terminal. This approaches fiber ubiquity and becomes very suitable to meet the anticipated needs of future dense mobile networks.

The phone, cellular and cable access networks provide significant opportunity to leverage their fiber resources already deployed or being deployed. Coherent optics can be leveraged to extract significant additional capacity from fiber resources already dedicated to other services. The efficiency of coherent optics enables effective reclaiming of fiber capacity when transitioning from analog sub-carrier modulation (SCM) and digital IM-DD optical links to coherent links. This creates fiber capacity at the edge of the network where a dense deployment of small cell sites through 5G is expected.

Coherent optics link margins in the access are significant due to its shorter fiber link lengths. In addition, the amount of dispersion that needs to be compensated is orders of magnitude lower than what regional or backbone networks have to compensate for. This simplifies the signal processing required for the access environment, and the link margin enables relaxation of component requirements.

REFERENCES

1. Statista.com, "Global digital population as of April 2019," 2019. www.statista.com/statistics/617136/digital-population-worldwide/
2. J. Nielsen, "Nielsen's law of internet bandwidth," January 6, 2018. www.nngroup.com/articles/law-of-bandwidth/
3. J. Karafin, B. Bevensee, "On the support of light field and holographic video display technology," *Society of Information Display SID Digest*, Volume 25, Issue 2, 2018, pp. 318–321.
4. International Telecommunications Union Recommendation ITU-T G.987.2, "10-Gigabit-capable passive optical networks (XG-PON): Physical media dependent (PMD) layer specification," February, 2016. www.itu.int/rec/T-REC-G.987.2-201602-I/en
5. International Telecommunications Union Recommendation ITU-T G.9807.1 "10-Gigabit-capable symmetric passive optical networks (XGS-PON): Access networks: Optical line systems for local and access networks," June, 2016. www.itu.int/rec/T-REC-G.9807.1-201606-I/en
6. IEEE 802.3av: Part3, Amendment 1: Physical Layer Specifications and Management parameters for 10 Gb/s Passive Optical Networks IEEE Standard, 2007.
7. C. Knittle, "Next generation PON 100G-EPON," IEEE 802, ITU-T SG15, January 27, 2018. www.itu.int/en/ITU-T/Workshops-and-Seminars/20180127/Documents/2.%20Curtis%20Knittle.pdf
8. NCTA – The Internet and Television Association, 25 Massachusetts Ave. NW #100, Washington DC, 2001. www.ncta.com/industry-data
9. PC Magazine Encyclopedia. Definition of central office, www.pcmag.com/encyclopedia/term/39518/central-office
10. T. Pfeifer, "Next generation mobile fronthaul and midhaul architectures," *IEEE/OSA Journal of Optical Communications and Networking*, Volume 7, Issue 11, November 1, pp. B38–B45, 2015.
11. J. Wang, Z. Jia, L.A. Campos, C. Knittle, "Delta-sigma modulation for next generation fronthaul interface," *Journal of Lightwave Technology*, Volume 37, Issue 12, June 15, pp. 2838–2850, 2018.
12. CTIA report, "The state of wireless 2018," 2018. https://api.ctia.org/wp-content/uploads/2018/07/CTIA_State-of-Wireless-2018_0710.pdf

2 Direct-Detection Systems for Fiber-Access Networks

Luis Alberto Campos, Junwen Zhang, and Mu Xu

CONTENTS

2.1 INTRODUCTION

Driven by mobile Internet, cloud networking and video-streaming services, the bandwidth requirements in transport networks, metro networks and short-distance optical networks have grown tremendously in recent years [1,2]. Considering the cost and complexity, intensity-modulation and direct-detection (IM-DD) have played significant roles in short-distance optical networks, including optical access networks, intra- and inter-data center interconnections [3–8]. It is envisioned that for optical-access networks, data rates per wavelength at 25, 50 Gb/s and even beyond will be required [9–12]. Recently, 100G, 400G and even beyond data transmissions, based on compact and low-cost transceivers, have been intensively studied in inter- and intra-data-center interconnections and metro networks [7,8], while for access network, IEEE 802.3ca Task Force is making progress toward the standardization of the next generation 25/50G Ethernet passive optical network (EPON) in O-band [9].

Moving toward higher data rates, direct detection with high-order modulation formats is a more practical and effective method in these systems [3–12], for which different techniques have been proposed based on direct detection, such as duobinary, pulse amplitude modulation (PAM) [6–7], discrete multi-tone (DMT) or orthogonal frequency division multiplexing (OFDM) [8], carrier-less amplitude/phase (CAP) modulation [13–15], and quadrature amplitude modulation (QAM) and subcarrier modulation (SCM) [16]. This chapter discusses the analog and digital IM-DD optical communication systems with different high-order modulation formats. The principles of generation, detection and signal recovery for PAM-4, CAP and DMT signals have also been discussed. Digital signal processing (DSP), which plays a significant role in these systems, has also been covered.

2.2   ANALOG OPTICS

Analog optics refers to the modulation of an optical carrier by radio frequency (RF) or microwave signals. This approach has been used for remoting the antenna in military applications and for carrying native RF cellular signals in front-haul applications. In addition, it is widely used in the cable space to deliver data and video services over frequency-multiplexed RF channels that intensity modulates an optical carrier. While the military and cellular use cases of analog optics have been narrowband for the most part, in the cable environment, it is a broadband multi-octave environment that requires special considerations.

Since the 1990s, cable has been delivering services over their hybrid fiber coaxial (HFC) networks when its significant infrastructure upgrade took place. In this HFC access network, 6–8 dedicated fibers connect the hub to a fiber node. Each hub typically connected to tens of fiber nodes. The signal, at the fiber node, transitions from optical-to-electrical domain and vice-versa.

In most cases, the RF signal aggregate of the entire downstream and all of its many frequency-multiplexed 6- or 8-MHz subcarriers directly modulates the current of a laser diode in order to carry the aggregate RF information to the fiber node. At the fiber node, a photo-diode detects the optical signal, generating the electrical signal that reproduces the original RF downstream signal generated at the hub. This signal is amplified and delivered through the coaxial branches that follow the fiber node.

2.2.1   CABLE LEGACY SIGNALS

The prevalent service through the early fiber distribution infrastructure in cable has been video services. In cable, since the aggregate of frequency multiplexed RF channels is used to modulate an optical carrier for transmission, it is also described as subcarrier multiplexing. The older analog video signals, depending on the region of the world where they originate, use National Television System Committee (NTSC),

Phase Alternating Line (PAL) or Sequential Couleur avec Memoire (SECAM) analog video formats [17,18]. In general, they consist of three modulated carriers. The strongest carrier is used to carry luminance information, the second carries audio information and the third one color. Transmission of these strong carriers are subject to intermodulation and other types of nonlinear distortion. In the United States, where NTSC signal format is used, analog video signals are transmitted 6-MHz apart, forming a spectrum channel line-up that included of up to 80 frequency-multiplexed 6-MHz analog video channels.

In order to transmit a multitude of these signals, side by side within the cable RF spectrum, the optical transmission systems carrying these signals require very high dynamic range and very low distortion to provide suitable transmission fidelity. Composite second order and triple beat distortion, and carrier to any discrete interference distortion requirements no greater than −53 dBc is mandated in industry specifications to support the transport of analog channels [19].

In their quest for greater transport efficiencies, the cable industry has been moving away from the transport of analog video channels to carry digital video channels. One 6-MHz channel would carry one NTSC video channel but could carry up to 12 standard definition digital video channels. The number of digital video channels reduces as one transitions to high-definition digital video transport.

As the number of channels dedicated to analog video in the channel line-up sanctioned lesser space to digital video transport and leveraged the transmission of the digital video at lower power levels than analog video, the distortion requirements for digital video transmission have been relaxed [20].

In North America, the 6-MHz-wide channels carry digital signals modulated using 64 and 256 QAMs [20]. Newer data over cable service interface specification (DOCSIS) transport system uses RF channels as wide as 192 MHz with up to 7,680 narrowband subcarriers per downstream channel. The modulations of these newer DOCSIS channels range from quadrature phase shift keying (QPSK) to 16,384 QAM [21]. A carrier-to-noise ratio (CNR) approaching 50 dB is estimated for transmissions using 16,384-QAM modulation.

The distortion as well as the CNR requirements on an analog optical link are very stringent. In order for a directly modulated laser (DML) to meet these linearity and noise requirements, it has to be operated at very high optical power levels. Many times, in order to achieve this performance level, the laser is operated with a transmit optical power exceeding 13 dBm.

The very high optical powers required to carry the high-quality RF signals and meet the high dynamic range limit the transport efficiency within a fiber strand. Nonlinear fiber conditions are present when optical power levels approach 20 dBm [22] in optical links less than 100 km. High-end analog lasers are capable to transmit at power levels of approximately 18 dBm. Adding multiple analog optics wavelengths requires careful positioning within the optical spectrum to limit nonlinear effects.

In the shorter access distances of analog optical links (<40 km), received optical power levels are high and the dominant source of noise is relative intensity noise or RIN. RIN dominates over shot noise and thermal noise at receive levels greater than 0 dBm. It also contributes to the limitation in signal-to-noise ratio (SNR) and dynamic range.

2.2.2 ANALOG OPTICS SOURCES

Typically, the direct modulation of Fabry–Perot lasers at lower cost don't meet these stringent linearity requirements. The downstream semiconductor lasers that meet these requirements and are suitable for high dynamic range applications are distributed feedback (DFB) lasers. One characteristic of DFBs is that they include a "corrugated" structure that imposes periodic boundary conditions on the signals in the active region restricting them to a generation of just a single mode and significantly suppressing all other modes.

The corrugated structure is implemented by periodically varying the index of refraction within or near the active region. This corrugated structure generates distributed periodic reflections. Nulls are imposed at these reflection or feedback locations, preventing the generation of multiple modes and only allowing transmission of the mode that matches the periodic structure.

DFB's single wavelength capabilities, along with carrier and light-confinement in structure designs of laser, lead to high photon densities and high optical transmit power levels. DFBs are extensively used in analog optics because they address the challenges in achieving high linearity and high optical power transmissions.

Optical chirp is another limiting factor introduced in the direct modulation process, particularly with the additional impact of fiber dispersion. The index of refraction in the laser active region changes with increase in current or carrier density. This change in the index of refraction causes chirp or an undesired frequency modulation that accompanies the desired intensity modulation of the optical carrier.

2.2.3 EXTERNAL MODULATION IN ANALOG OPTICS

External modulation is also used to implement analog optical links. As the laser current remains unchanged with external modulation, externally modulated links do not exhibit chirp. Externally modulated links using Mach–Zehnder modulators (MZMs) have been used in longer links. MZMs have a cosine-squared voltage to optical intensity response. The MZMs are biased at the optical intensity mid-point that enjoys maximum linearity. These cannot be driven to reach a maximum optical intensity swing unless linearization mechanisms are introduced.

The advantage of using an MZM is that the modulation function is decoupled from the optical source that can now be optimized for high power, higher speed, low noise and low intermodulation distortion. In comparison with direct modulation alternatives, the drawbacks faced while using MZM-based systems are size and cost. As photonic integration continues to evolve, this difference is less significant.

2.2.4 CABLE ACCESS NETWORK CHARACTERISTICS

Cable spectrum resources are divided in downstream and upstream portions. In North America, the downstream typically consists of a spectrum starting at a lower band edge of 50 MHz to an upper band edge of 750 MHz, 870 MHz, 1 GHz, or even 1.2 GHz, depending on the capabilities of the fiber node and amplifiers deployed. The upstream covers from 5 to 42 MHz, although higher frequency downstream/upstream splits are also supported. In Europe, the downstream channels are 8-MHz wide, and upstream/downstream transmission frequencies also differ. The downstream lower frequency edge is 85 MHz and the upstream upper band edge is 65 MHz.

In the fiber distribution network of cable, the hub-to-node distances are less than 40 km for about 90% of the optical links. Many of these short optical link distances can be addressed without in-line amplification. Erbium-doped fiber amplifier (EDFA) amplification is used in the longer links present in cable networks.

The popularity in fiber-connectivity services has generated fiber-strand shortages in high-demand areas which has typically been addressed using Wavelength division multiplexing (WDM). Wavelength division multiplexers in cable may only cover a portion of the C-band and may not necessarily exhibit the flat spectrum response that multiplexers in long-haul networks have.

In 20% of the access link cases, there is only one fiber strand available from hub to node. This means that bidirectional systems have to be designed for the single-fiber environment. To meet the high CNR and stringent distortion requirements of analog optics, the cable industry has extensively deployed Angle-Polished connectors. The cable-access environment has been designed to have negligible reflections, which has facilitated the introduction of single-wavelength–single-fiber simultaneous bidirectional coherent optics transmission discussed in Chapter 7.

2.2.5 Upstream Transmission

Except for the longer optical links, upstream transmission in cable also leveraged analog optics. In longer link scenarios, the upstream RF cable spectrum is digitized at the fiber node and is carried in digital form to the hub where it is converted back into RF domain.

Earlier shorter analog upstream links used Fabry–Perot lasers, but they were frequently overdriven, resulting in signal clipping both at the threshold current as well as at high current levels when lasers were pushed beyond their linear operating range. This has prompted the migration toward DFB lasers, which provided greater dynamic range and limited clipping events.

2.2.6 Analog Fiber to the Home

In addition to connecting hubs to nodes through optical fibers, cable operators have also deployed fiber-to-the-home (FTTH) networks to deliver cable's RF spectrum to homes leveraging analog optics. FTTH networks are mostly deployed in green field scenarios. This approach is called RF over Glass or RFoG, in which you have fiber coming from the hub, splitting to connect 32 or 64 subscribers. One RFoG challenge in the upstream is that multiple simultaneous optical transmissions could be in wavelength proximity such that the difference in wavelength would fall within the frequency response of the optical receivers generating optical beat interference. This can be solved through different means of wavelength control but would add complexity to the customer's premise device. Transmitters at the edge of the network have to be as simple as possible. While the downstream link in RFoG mirrors the challenges of traditional downstream transmission, there are specific shortcomings in upstream transmission of RFoG networks, including optical beat interference, laser clipping and dispersion, among others.

2.2.7 Aging and Temperature Control

Laser diodes used in analog optics operate at power levels that are higher than digital IM-DD systems and much higher that coherent optical systems. The higher optical power levels of higher currents accelerate the laser-aging process. The laser facets and the active region lattice structure degrade with optical intensity. The threshold current increases as the laser diode ages. Controlling the biasing of the laser diode is important to operate it in the optimal linear region.

The laser diode wavelength is dependent on temperature. Depending on the characteristics and design of laser diode, the wavelength of a laser may increase with temperature from a fraction of 1 nm/K to around 3 nm/K [23]. Introduction of temperature-control mechanisms are typically avoided to keep system complexity and cost low.

2.2.8 Distributed Access Architectures

The evolution of DOCSIS has led to the potential transport of 16,384-QAM signals in the downstream and 4,096 QAM in the upstream. This requires an optical transport link supporting a CNR approaching 50 dB. This level of performance is challenging to achieve through analog optics links. Analog optical transmit power levels, greater than 13 dBm, are needed to achieve this CNR level of performance. Such higher power levels limit the number of optical carriers that can be aggregated on a fiber without being impacted by fiber nonlinear distortion. Multiplexing more than eight analog optics signals at these higher modulation orders on a single fiber is not practical. An optimization exercise in a number of optical carriers, optical transmit power levels, placement of optical carriers at lower distortion wavelengths and modulation order of the DOCSIS RF payload has to be conducted. This is one of the reasons that has prompted the cable industry to move toward distributed architectures. In distributed architectures, the function of the cable modem termination system (CMTS) is split. Some functions remain at the hub location, while some are shifted to the edge. In one functional split scenario, physical layer (PHY) functionality is shifted toward the edge, which means that RF is generated locally at the node, media access control (MAC), an upper layer functions provided by the CMTS core or converged cable access platform (CCAP) core at the hub or a central location. Baseband digital optics is used to establish communication between the core components and the edge components at the fiber node. This change in architecture also transitions the optical access transport from analog optics to baseband digital optics.

2.3 PULSE AMPLITUDE MODULATION (PAM) SIGNALS

The complexity, performance, system setup and signal processing for different modulation formats are quite different [3]. Among the modulation formats mentioned above, PAM signals, especially PAM-4 signal, are one of the most attractive formats for short-distance optical communications in terms of simply signal generation and processing. For PAM signals, the data or information is encoded in the amplitude of a series of signal pulses. For instance, in PAM-4 signals, there are four possible discrete pulse amplitudes with an efficiency of 2 bits per symbol, while in PAM-8, there are eight possible discrete pulse amplitudes with an efficiency of 3 bits per symbol. Recently, PAM-4 signal has been standardized as the modulation formats for short-distance optical networking technologies, including the 200GBASE-SR4/DR4 in 802.3cd and 400GBASE-DR4/FR8/LR8 in IEEE802.3bs.

Recent technology advances for PAM signals are discussed in the following sections: theory and principle of PAM-4 signal generation and detection in optical communication systems, commonly used equalization and DSP for PAM-4 signals to overcome channel linear as well as nonlinear impairments and recent progress on other multi-level signals.

2.3.1 OPTICAL PAM-4 SIGNAL MODULATION, DETECTION AND EQUALIZATION

Electrical PAM signals can be generated by either signal multiplexing or digital-to-analog conversion (DAC). For instance, a PAM-4 signal with four electrical levels can be generated by multiplexing two paths of independent non-return-to-zero (NRZ) signals, in which one path is attenuated by 6 dB [24]. More advanced PAM signals, such as PAM-8 and PAM-16, can also be generated by three or four paths of independent NRZ signals [25]. As an example, in [24], the four-level 120-GBaud PAM-4 signal is generated by an electrical combiner from two de-correlated 120-Gb/s NRZ signals, and one of the NRZ signals is first reduced to half of the amplitude by a 6-dB attenuator.

Recently, with the maturity and development of high-speed ADC/DAC technologies, PAM signals can be simply generated by a DAC in which DSP can be enabled to improve the transmitter signal quality. DAC provides a more flexible way to generate any PAM signal level, with arbitrary pulse shaping and transmitter-side pre-compensation [26–29]. There are an increasing number of reports based on DAC for PAM signal generations. Moreover, 100 GBaud DAC-based PAM-4 signal generation is also available [29].

For signal modulation, several 100 Gb/s per lane short-reach transmission experiments have been reported recently, based on PAM with both integrated or separated lasers and modulators [26–29]. As might be expected, compared with external modulation based on MZM [26,27], relatively simple in implementation IM-DD schemes, i.e., PAM-4, and low-cost, low-form factor transmitters, such as electro-absorption modulated lasers [7] or DMLs [28], would be promising technologies for future inter-data center interconnection applications. In some applications in which huge quantity of power budget is required, semiconductor optical amplifiers (SOA) can be used to boost output power, as reported in [12]. High-bandwidth direct-modulation modulators are also under extensive studies, as the most recently reported silicon–organic hybrid modulator achieved 200G per wavelength PAM-4 modulations [29].

At the receiver's side, positive–intrinsic–negative (PIN)-type photodiode and avalanche photodiode (APDs) are widely used in direct-detection optical system to convert optical data into electrical form. In general, APD has a greater level of sensitivity as compared with PIN, which can be used for longer fiber-transmission networks and higher power budget links. The avalanche action in APD greatly increases the gain of the diode by many times, achieving much higher sensitivity [12]. However, an APD requires a higher operating voltage to drive it. In some direct-detection systems, SOA or EDFA are also used for pre-amplification before the photodiode. To suppress the out-of-band (OOB) noise, an optical filter is generally used at the receiver's side after the optical pre-amplification. In general, the noise type and distribution for different receiver type is different. For instance, amplified spontaneous emission noise is the major one in an SOA-based system, while the dominant source is the shot noise in APD-based receiver and thermal noise contributes the most for PIN receivers [30].

2.3.2 DIGITAL SIGNAL-PROCESSING METHODS FOR PAM SIGNALS

Linear equalization and nonlinear compensation are widely used for PAM signals. Figure 2.1 shows the transmitter and receiver-side signal processing flows for PAM signals. After PAM-4 symbol mapping, one time-domain finite impulse response (FIR) filter can be used to pre-compensate the channel response and improve the signal quality. The channel response can be measured off-line or online by a broadband receiver [7,12,30]. In some reports, nonlinear compensations are also reported to equalize the pulse levels in the system [7], which can be implemented right after symbol mapping. Since PAMs are multi-level signals, the nonlinear compensation can be very simple by using a look-up-table–based method as reported in [7]. After the transmitter-side DSP, the PAM signals are resampled and converted to analog signal by DAC.

At the receiver's side, first the sampled signal is resampled to one or two samples per symbol and then processed by signal and clock recovery algorithms – for instance, digital clock recovery and adaptive channel equalization. Gardner re-timing algorithm and Square timing recovery are reported [12] for digital time recovery.

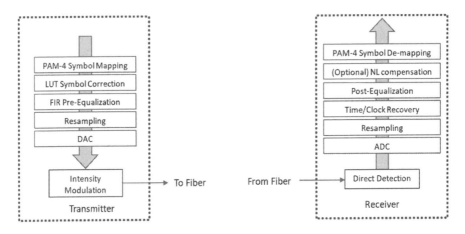

FIGURE 2.1 The transmitter and receiver-side signal processing flows for PAM signals.

Linear equalization based on feed-forward equalizer or decision feedback equalizer is also commonly used for receiver-side post equalizations [12].

To support 400G data connection, one of the most promising solutions is 4-lane × 100 Gb/s/λ net rate based on PAM signals, when considering the transceiver complexity, size, power consumption and cost [7,27]. However, the transmission distance in these reports is limited to about 10 km due to the frequency-related power fading induced by chromatic dispersion (CD) after direct detection in C-band. Inline optical CD compensation is used in some reports for the 80-km links [7]. As might be expected, systems without inline optical CD-compensation modules would be simpler and more flexible. Although single-sideband (SSB) modulation or optical filtered vestigial sideband modulation can achieve power fading-free spectrum after direct detection, they have 3-dB Optical Signal to Noise Ratio (OSNR) penalty compared with double-sideband (DSB) signals [31]. Other solutions, like Stokes vector direct detection or block-wise phase shift direct detection, require complicated receiver-setup like coherent detection, which deviates from the original intention of low cost [32]. Square-law direct-detection prohibits the post-CD compensation under severe power-fading; therefore, CD pre-compensation is considered a more effective method [31]. However, the nonlinear (NL) impairments caused by modulators and the interplays between residual CD and direct-detection process further degrade system performance [32,33].

In [33], the authors proposed and demonstrated a direct-detection transmission scheme for PAM-4 signals using digital CD pre-compensation and advanced NL-distortion compensations, achieving four lanes of 112-Gb/s PAM-4 signals more than 400 km without any optical CD compensation. Figure 2.2 shows the principle of digital CD pre-compensation, in which an optical in-phase and quadrature (I/Q) modulator is used for digital pre-compensated PAM signals' modulation. The CD pre-compensation is based on the channel response of fiber dispersion, implemented in the frequency domain. Since the CD pre-compensation includes phase information, the I/Q modulator is, therefore, driven by independent I and Q waveforms after CD pre-compensation. The authors in [34] also demonstrated that dual-arm MZM can also be used for CD pre-compensation, in which the two arms are used for I and Q modulations. In these reports, they've also shown that additional post–NL distortion compensation, i.e., Volterra nonlinear compensation can further improve the performances.

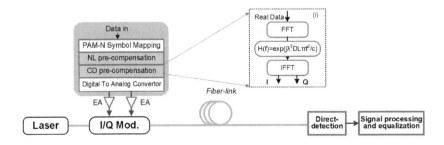

FIGURE 2.2   The principle for CD pre-compensation and NL pre-compensation.

2.3.3 CHALLENGES AND RECENT PROGRESS ON HIGH-LEVEL PAM SIGNALS

Higher order modulation formats or higher level PAM signals are achieving acceptability as they can carry more data with the same signal baud rate. Recently, there are many reports on PAM-8, as well as on PAM-16 signals in direct-detection system. However, there are many challenges in these systems. On the one hand, higher order will introduce large power penalty [35]. Compared with PAM-4 signals, theoretical study shows about 3.6-dB power penalty for PAM-8 and additional 3.3-dB penalty for PAM-16 signal. To achieve flexible data rate, spectrum efficiency and smoothen the power step–like penalty between different PAM orders, probabilistically shaped coded modulation is also proposed for PAM-based IM-DD system [36].

On the other hand, nonlinear impairment will become more significant as high-level signals will require highly linearity for signal generation, modulation and detection. Therefore, recently, there are many studies on nonlinear impairments compensation. NL-maximum likelihood sequence estimation based on Volterra Filter is used to achieve a 255-Gbps PAM-8 Transmission IM-DD link for receiver-side post compensation [37], while a combined use of Tx-side look up table–based amplitude and Rx-side eye-skew compensation is also demonstrated to improve the performance of PAM-8 signals in [38].

2.4 CARRIER-LESS AMPLITUDE/PHASE (CAP) MODULATION

CAP modulation is a variant of QAM [4,13]. The QAM signal in CAP is generated by combining two PAM signals filtered through two orthogonal impulse responses filter pair, and, in general, from a Hilbert pair [13]. CAP is first used in asymmetric digital subscriber line (ADSL) systems and has recently been introduced to short-reach optical interconnections, as it is believed to have reasonable implementation complexity and good performance. For instance, it allows relatively high data using low-cost optical components, such as DML and vertical cavity surface-emitting laser, and optical and electrical components of limited bandwidth [13]. Compared with OFDM [8] and QAM–SCM [16], the complexity is reduced as no electrical complex-to-real-value conversion, complex mixer, RF source or optical in-phase/quadrature (I/Q) modulator are required for CAP. It does not require the discrete Fourier transform that is utilized in OFDM signal generation and demodulation [8]. A number of optical communication systems based on CAP have been demonstrated recently [13–15]. In [14], multi-band CAP-16-QAM has been proposed to extend the bandwidth for high-speed short-range data transmission. In [13], the digital equalizer based on cascaded multi-modulus algorithm (CMMA) has been proposed for CAP-16-QAM inter-symbol interference (ISI) equalization with good performance. The first high-level modulation format CAP system, CAP-64-QAM with 60-Gb/s data rate is demonstrated in [39], shows potential applications in high-speed optical communications.

2.4.1 THE PRINCIPLES OF SINGLE-BAND CAP

The schematic diagrams of transmitter and receiver based on CAP m-QAM for fiber-wireless transmission are shown in Figure 2.3. The two-dimensional CAP can be generated by using two orthogonal filters as the filter pair [13,39]. The original bit sequence is first mapped into complex symbols of m-QAM (m is the order of QAM); then the mapped symbols are up-sampled to match the sample rate of shaping filters. The sample rate of shaping filters is determined by the data baud and DAC sample rates. For CAP generation, the I and Q components of the up-sampled sequence are separated and sent into the digital shaping filters, respectively. The outputs of the filters are subtracted to be combined together as $S(t)$ for optical modulations after DAC. At the receiver's side, the received signal after down-conversion is fed into two matched filters to separate the I and Q components.

The orthogonal and matched filter pairs $f_I(t)$, $f_Q(t)$, $mf_I(t)$ and $mf_Q(t)$ are the corresponding shaping filters and form a so-called Hilbert pair in the transmitter and receiver. The two orthogonal filters are constructed by multiplying a square root–raised cosine pulse with a sine and cosine function, respectively, as described in [13]. At the receiver, generally, we have the matched filters with relations as $mf_I^n(t) = f_I^n(-t)$ and $mf_Q^n(t) = f_Q^n(-t)$. In this way, for the CAP receiver, the I and Q data after matching the filter pair can be obtained as $R(t)$, which is the CAP signal after down-conversion at the receiver, and $r_I(t)$ and $r_Q(t)$ are the output after matched filter pair. After down-sampling, a linear equalizer is employed for the complex signal, and a decoder is utilized to obtain the original bit sequence.

The CAP signal can be generated with multi-bands, which will further improve the overall capacity [13–15,40]. Different from the single band CAP in [14], multi-band CAP uses multi-filter pairs that is located in different frequency sub-bands [40]. For each sub-band, the two orthogonal filters are constructed by multiplying a square

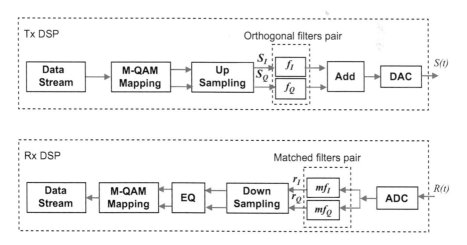

FIGURE 2.3 Schematic diagrams of transmitter and receiver based on CAP m-QAM for CAP signal generation and processing.

root–raised cosine pulse with a sine and cosine function, respectively. The center frequency of cosine and sine function determines the sub-band center frequency. Recent reports show that the multi-band CAP signal can be used for sub-band multiplexing to achieve high-data rate [39] and also for multi-user fiber-access networks [40].

2.4.2 DIGITAL SIGNAL PROCESSING FOR CAP SIGNAL IN SHORT-REACH OPTICAL NETWORK

Since the appropriate sampling time is hard to decide, the sampling time offsets will lead to subsequent signals seriously affected by ISI and the crosstalk between the in-phase and quadrature components. One-phase rotation is also produced by the crosstalk. Therefore, after down-sampling, a linear equalizer is employed for the complex signal, and a decoder is utilized to obtain the original bit sequence. In the system, the orthogonal and matched filters are realized by digital FIR filters with a tap length of transmitter orthogonal filter length and receiver matched filter length, respectively. As analyzed in [4], the tap length of the FIR filters determines the filter shape of time domain pulse and frequency response. In the experiment, the impact of tap length of these FIR filters on the system performance is studied.

In [14], the authors use two-stage equalizations, ISI equalization and phase recovery (PR), to equalize the CAP signal. The ISI equalization is performed using a constant modulus algorithm (CMA) for pre-convergence followed by the CMMA. However, it is difficult for high-order CAP-QAM signal equalization using CMMA, since the ring spacing in QAM is generally smaller than the minimum symbol spacing. It has been proved that DD-least mean square (LMS) can achieve better SNR performance compared with CMMA for high-order QAMs [39]. However, since CMMA is based on the radius of the circles of the symbol, it is a phase-independent algorithm. In this way, additional PR is required following the CMMA to equalize the crosstalk. Therefore, a cascaded two-stage equalizer based on DD-LMS after CMA pre-convergence for CAP-QAM signal ISI and crosstalk equalization is proposed and demonstrated in [39]. In general, the commonly used equalization methods for PAM and QAM can also be used for CAP signals.

2.5 DISCRETE MULTI-TONE IN OPTICAL ACCESS NETWORK

DMT is a kind of multi-carrier modulation format based on OFDM. In regular OFDM, each subcarrier carries a complex-valued QAM symbol, which results in multiple orthogonal signal components with in-phase (I) and quadrature (Q) tributaries on the RF band. OFDM has become one of the most frequently used data formats for today's wireless data transmission systems. For example, a typical Long-Term Evolution (LTE) OFDM component carrier occupies a bandwidth of around 20 MHz on a RF carrier with 2,048 total subcarriers and 1,201 loaded subcarriers. Being similar to the OFDM, DMT is also generated through inverse fast-Fourier transform (IFFT) and fast-Fourier transform (FFT). The major difference between OFDM and DMT is that the signal generated through DMT is real-valued in the signal's amplitude domain, which does not contain any phase information. OFDM and DMT have been used in a large scale for various telecommunication and data-transmission standards, such as

IEEE 802.11a/g/n/ac/ad wireless local access network, IEEE 802.15.3a wireless personal area network, IEEE 802.16 worldwide interoperability for microwave access, digital video broadcasting-terrestrial, LTE under third-generation partnership project, high-speed digital subscriber lines, ADSLs, and DOCSIS [41].

In an optical system, DMT signal can be transmitted through an IM-DD link, in which the transmitter could be a DML or an MZM plus a laser source, and the receiver typically comprises a photo detector. The signal generation and reception procedures are shown in Figure 2.4. The first step is to map the binary information into QAM symbols. The symbol streams will then be converted from serial into parallel, where each symbol in one column of the paralleled data is carried by one subcarrier. IFFT is applied to transform the signals from frequency to time domain. Then, a part of the samples at the end of each symbol after IFFT is replicated as a cyclic prefix (CP), which will be added to the beginning of the time-domain signals before they are converted back to a serial of data samples. In DMT, the signals have to be real-valued. There are two approaches to achieve this: first, in the frequency domain before IFFT, only half of the subcarriers are loaded with QAM symbols, while the other half is generalized by reversing the order and taking the conjugate of the first half. In another approach, after loading one half of the subcarriers in the frequency domain, IFFT is applied and CP is added. Then only the real part of the signals is retained before it is sent to the following steps. A digital-to-analog converter (DAC) is used to convert the digitized samples into analog ones, which are sent to the transmitter and modulated onto the light. The signal recovery process at the receiver site is basically a reversed process compared with the transmitter site, and the signal flows will pass through functioning blocks as analog-to-digital converter (ADC), serial-to-parallel converter, CP removal, FFT, equalizer and symbol decoder. However, since the signal is real, which means that half of the frequency components is simply the conjugate of another half. Thus, after FFT, only one half of the double-sided band signal will be considered for equalization. Since DMT has distributed the signals onto many subcarriers and the response of each subcarrier can be regarded as a flat function across a narrow bandwidth, a simple one-tap equalizer can be used for equalizing the overall channel response in the frequency domain. Such a one-tap equalizer is much more computationally efficient than the LMS algorithm–based adaptive digital filters in processing single-carrier signals.

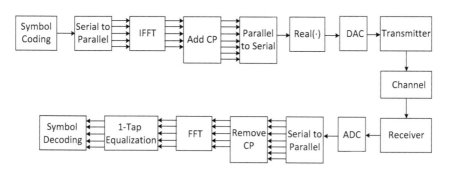

FIGURE 2.4 Digital signal recovery procedures for discrete multi-tone (DMT).

Since the DMT signal comprises a group of subcarriers, the power and modulation format of each subcarrier can be reconfigured to be adaptive to the channel response, which maximizes the bandwidth efficiency or minimizes the power consumptions. The basic idea to improve the bandwidth efficiency is to assign higher order and lower order QAM signals to the subcarriers with higher and lower SNR, respectively. Water-filling [42] and finite-granularity loading algorithms [43] are the two commonly used methods in bit loading of DMT systems. The former is more precise, and the latter is more computational efficient. Power-loading algorithm to minimize the power consumption based on Lagrange-multiplier algorithm is also proposed in [44], which is also a frequently used algorithm for optimizing the power or information allocation on the subcarrier domain.

OFDM or DMT with CP has dominated the modulation format in a number of communication or data-transmission systems for a long time. However, whether OFDM is the optimal format for wireless and wired access network has incurred a lot of debate due to some of its intrinsic issues. First, a plain OFDM transmission with a rectangular window to perform IFFT and FFT has very strong OOB power leakages. When aggregating multiple-component carriers on the same frequency band, guard bands have to be inserted between adjacent component carriers to avoid inter-channel interference resulted by the OOB power leakage. These guard bands significantly reduce the spectral efficiency of the system by around 30%. Meanwhile, OFDM requires CP to protect it from ISI caused by multi-path effects. The spectral efficiency will be further reduced by CP around 5%–20% depending on the number of additional CP samples needed to overcome the channel delay spread across all subcarriers. Recently, some new multi-carrier modulation formats have also been studied, including filter-bank multi-carrier [45], universal-filtered multi-carrier [46] and generalized frequency-division multiplexing [47]. They all apply some particular filtering techniques to shape each subcarrier or a bundle of subcarriers to suppress the OOB power leakage, which trades off some computational complexities for improved bandwidth efficiencies.

2.6 ADVANCED DIRECT-DETECTION TECHNOLOGY BASED ON KRAMERS–KRÖNIG RECEIVERS

Dual-polarization coherent optical-communication technology could maximize the spectral efficiency of fiber-transmission system since it encodes the information on both quadratures and polarizations of the electrical field, which empowers coherent optics as one of the major solutions for today's long-haul and metro networks. However, there are also some challenges to overcome. First, when multiplexing data into four orthogonal channels, the complexity of coherent transmitter and receiver is significantly higher than IM-DD systems, leading to significant increase in cost. Second, to guarantee the orthogonality of the four channels, the phase, polarizations, clocks and skews need to be carefully controlled, which further increase the complexity and operational expenditure of the system. These challenges are major obstacles limiting coherent technologies from further penetrating into the short-haul and access networks.

Recently, some advanced direct-detection technologies have been developed to compete with coherent optics. They trade in some of the spectral efficiency for simplifications of the system architecture, reduced cost and improved receiver sensitivity.

This provides a good transient solution to increase the capacity of short-haul and access networks. One of the promising approaches is self-coherent Kramers–Krönig (KK) method [48]. KK-based direct-detection scheme leverages the reconstruction of amplitude and phase in minimum phase signals using the KK relations. Given a complex signal, $s(t)$, which occupies a bandwidth of B, distributed from $-B/2$ to $B/2$, we can obtain the SSB signal as $h(t) = A * \exp(j\pi Bt) + s(t)$, in which the trace of the signal can be plotted over a complex plane with real and imaginary axes as shown in Figure 2.5. Then it is said that $h(t)$ fulfills the minimum phase criterion if its trajectory distribution does not encircle the origin as shown in Figure 2.5(b). It can also be proved that minimal phase criterion is sufficiently satisfied when $|A| > |s(t)|$, in which the distribution of the signal's trajectory is plotted in Figure 2.5(b). When minimal phase property is met, the complex signal can be perfectly reconstructed from its field-intensity distribution after power-law detection, $I(t) = |h(t)|^2$. However, if the signal does not fulfill the minimal phase criterion, it will result in phase ambiguity, especially for the part with the trajectory passing across the zero point. Thus, the complex signal cannot be reconstructed correctly. Two equations are essential for rebuilding the amplitude and phase of the complex signal from its field intensity [49], which are given as

$$\varphi_s(t) = \frac{1}{2\pi} \, p.v. \int_{-\infty}^{\infty} dt' \, \frac{\log[I(t)]}{t - t'} \tag{2.1}$$

$$S(t) = \left\{ \sqrt{I(t)} \exp[i\varphi_s(t)] - A \right\} \exp(i\pi Bt) \tag{2.2}$$

These equations are also referred to as KK relations. The system architectures between coherent and KK detector-based optical links are compared in Figure 2.6(a) and (b), respectively. Figure 2.6(b) shows that the hardware of KK system is greatly simplified, in which an I/Q modulator is utilized as the transmitter and a simple photo detector is integrated with an analog–digital converter plus digital-signal-processing chips comprises the receiver part. For the signal generation at the transmitter site, except from the upconverted complex signal, one frequency tone will be added. This frequency tone acting as a local oscillator (LO) has to be inserted at the edge of a bandwidth limited signal, and its strength needs to be carefully controlled. A too weak LO cannot fully guarantee a satisfactory minimum phase condition. However, a too

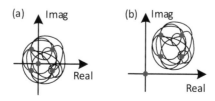

FIGURE 2.5 Trajectories of signals when (a) minimum phase criterion is not satisfied and (b) minimum phase criterion is satisfied.

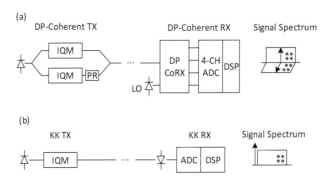

FIGURE 2.6 System architectures of (a) regular coherent optical system and (b) Kramers–Krönig-based optical receiver.

strong LO may consume too many carriers at the photo detector, thus weakening the receiver sensitivity of the system. One of the major benefits of KK receiver lies in that the system is robust against nonlinear distortions. In traditional optical SSB system, the signal-to-signal-beat interference (SSBI) significantly reduces the signal quality, and the digital-signal-processing algorithms to eliminate the SSBI are quite complex or unreliable [50,51]. The method to avoid SSBI based on adding guard bands between the DC and signal's components wastes too much bandwidth of the system [52]. Nevertheless, in KK system, the signal recovery is based on the relations between amplitude and phase in Equations (2.1) and (2.2), which is free from the SSBI, as a second-order nonlinear effect. With this unique benefit to suppress signal–signal beat noise term, notable improvements in receiver sensitivity can be achieved for direct-detection systems. However, one of the major disadvantages of KK receiver is that the logarithm operation in Equation (2.1) broadens the spectrum of the signals [48]. So, two to four times oversampling is typically required for signal recovery in the KK system, which remarkably increases the cost for analog-to-digital converters.

Recently, with the growing interests in KK receiver–based advanced direct-detection system, a lot of innovative works and interesting results have been published. A polarization multiplexed KK receiver has been demonstrated in [53], which doubles the system spectral efficiency. A modified KK receiver to accurately reconstruct the field at low carrier-to-signal-power ratio has also been proposed [54]. By applying exponential operation to an SSB signal, the minimal phase criterion is automatically satisfied, which greatly simplifies the signal-generation process. However, the system may also suffer from increased peak-to-average-power ratio incurred by the exponential operation. A twin SSB KK receiver is also proposed in [55], like traditional twin SSB scheme [56], with two carefully configured bandpass optical filters, one DSB optical signal with two quadratures can be filtered into two SSB signals, where both of them satisfy the minimum phase condition. Thus, the traditional KK receiver is extended to recovering any kinds of DSB coherent optical signals with in-phase and quadrature components. However, the new architecture also greatly increases the system's complexity, which incurs stringent requirement for the optical filter design with a sharp roll-off factor and precise tunability. In general, although it could improve the receiver's sensitivities of the direct-detection system, the KK receiver still faces a lot of technical issues, including maintaining the minimum phase condition of the signals, increased over-sampling rate and high complexity in digital-signal-processing part.

2.7 SUMMARY

In this chapter, analog optics and baseband IM-DD digital optics are introduced, including direction detection with different high-order modulation formats aided by DSP.

In analog optics, the transition from analog-to-digital video was discussed as well as the evolution from single 6- or 8-MHz wide 256-QAM channels to 192-MHz-wide channels with subcarriers modulated up to 16,384 QAM. Use of FTTH RFoG networks in green field scenarios and the transition from a centralized architecture to a distributed architecture with RF generation at the node are also reviewed.

In digital optics, the principles of generation, detection and signal recovery for PAM-4, CAP and DMT signals are explained. In higher speed IM-DD system, DSP becomes more and more essential as these high-order modulation formats suffers both channel linear and nonlinear impairments. Analysis and comparison of complexity, performance and cost for these modulation formats can be found in [4,57]. Overall, PAM signals show less complexity, simple implementation and low-power consumption for DSP. Therefore, PAM-4 has been widely used and standardized in different optical systems. However, we also observed that 200G, 400G and beyond per channel will be more and more challenging for IM-DD systems, even that these high-order modulation formats are used. Therefore, coherent optics will play significant roles in future high-speed short-reach optical networks.

REFERENCES

1. R. P. Davey and D. B. Payne, "The future of optical transmission in access and metro networks – An operator's view," 2005 31st European Conference on Optical Communication, ECOC 2005, vol. 5, Symposium We 2.1.3, 2005. DOI:10.1049/cp:20050835

2. J. S. Wey and J. Zhang, "Passive optical networks for 5G transport: Technology and standards," *Journal of Lightwave Technology*, vol. 37, no. 12, pp. 2830–2837, 2019.

3. J. Lee, N. Kaneda, T. Pfau, A. Konczykowska, F. Jorge, J.-Y. Dupuy, and Y.-K. Chen, "Serial 103.125-Gb/s transmission over 1 km SSMF for low-cost, short-reach optical interconnects," Proceeding Optical Fiber Communication Conference (OFC), 2014, paper Th5A.5.

4. K. Zhong, X. Zhou, T. Gui, L. Tao, Y. Gao, W. Chen, J. Man, L. Zeng, A. P. T. Lau, and C. Lu, "Experimental study of PAM-4, CAP-16, and DMT for 100 Gb/s short reach optical transmission systems," *Optics Express*, vol. 23, pp. 1176–1189, 2015.

5. A. Dochhan, N. Eiselt, H. Griesser, M. Eiselt, J. J. V. Olmos, I. T. Monroy, and J.-P. Elbers, "Solutions for 400 Gbit/s inter data center WDM transmission," European Conference on Optical Communication, pp. 680–682, 2016.

6. Y. Gao, "112 Gb/s PAM-4 using a directly modulated laser with linear pre-compensation and nonlinear post-compensation," European Conference on Optical Communication, pp. 121–123, 2016.

7. J. Zhang, J. Yu, and H.-C. Chien, "EML-based IM/DD 400G (4×112.5-Gbit/s) PAM-4 over 80 km SSMF based on linear pre-equalization and nonlinear LUT pre-distortion for inter-DCI applications," 2017 Optical Fiber Communications Conference and Exhibition (OFC), Los Angeles, CA, 2017, paper W4I.4.

8. Y. Wang, "Demonstration of 4x128-Gb/s DFT-s OFDM signal transmission over 320-km SMF with IM/DD," *PJ*, vol. 8, no. 2, pp. 1–9, 2016.

9. V. Houtsma, D. van Veen, E. Harstead, "Recent progress on standardization of next generation 25, 50 and 100G EPON," *Journal of Lightwave Technology*, vol. 35, no. 5, pp. 1228–1234, 2017.

10. ITU-T Q2/SG15, "Proposal for the study of 50G TDM-PON," ITU-T SG 15, Contribution, vol. 641, 2017.

11. M. Tao, L. Zhou, H. Zeng, S. Li, and X. Liu, "50-Gb/s/λ TDM-PON based on 10G DML and 10G APD supporting PR10 link loss budget after 20-km downstream transmission in the O-band," Optical Fiber Communication Conference, p. 2, 2017, paper Tu3G.2.

12. J. Zhang, J. S. Wey, J. Yu, Z. Tu, B. Yang, W. Yang, Y. Guo, X. Huang, and Z. Ma, "Symmetrical 50-Gb/s/λ PAM-4 TDM-PON in O-band with DSP and Semiconductor Optical Amplifier Supporting PR-30 Link Loss Budget," Optical Fiber Communication Conference, OSA Technical Digest, Optical Society of America, 2018, paper M1B.4.

13. L. Tao, Y. Wang, Y. Gao, A. P. T. Lau, N. Chi, and C. Lu, "40 Gb/s CAP32 system with DD-LMS equalizer for short reach optical transmissions," *IEEE Photonics Technology Letters*, vol. 25, no. 23, pp. 2346–2349, 2013.

14. M. Iglesias Olmedo, Z. Tianjian, J. Bevensee Jensen, Z. Qiwen, X. Xu, and I. T. Monroy, "Towards 400GBASE 4-lane solution using direct detection of MultiCAP signal in 14 GHz bandwidth per lane," Presented at the Optical Fiber Communication Conference/ National Fiber Optic Engineers Conference, Anaheim, CA, 2013, paper PDP5C.10.

15. J. Wei, Q. Cheng, D. G. Cunningham, R. V. Penty, and I. H. White, "100-Gb/s hybrid multiband CAP/QAM signal transmission over a single wavelength," *Journal of Lightwave Technology*, vol. 33, pp. 415–423, 2015.

16. F. Zhang, K. Zou, and Y. Zhu, "High capacity optical transmission with nyquist sub-carrier modulation and direct detection," Advanced Photonics (IPR, NOMA, Sensors, Networks, SPPCom, SOF), OSA Technical Digest, Optical Society of America, 2016, paper SpW3F.6.

17. International Telecommunications Union Recommendation ITU-R BT.1700, "Characteristics of composite video signals for conventional analogue television systems," www.itu.int/rec/R-REC-BT.1700/en, February 2005.

18. International Telecommunications Union Recommendation ITU-R BT.1701-1, "Characteristics of radiated signals of conventional analogue television systems," www. itu.int/dms_pubrec/itu-r/rec/bt/R-REC-BT.1701-1-200508-I!!PDF-E.pdf, August 2005.

19. "Digital cable network interface standard," ANSI/SCTE 40, 2011, www.scte.org/ documents/pdf/standards/SCTE_40_2011.pdf

20. "Downstream RF interface specification – CM-SP-DRFI-I16–170111," 11 January 2017, Cable Television Laboratories, Inc.

21. "DOCSIS 3.1 Physical layer specification – CM-SP-PHYv3.1-I09–1606021," 21 June 2016, Cable Television Laboratories, Inc.

22. S. P. Singh and N. Singh, "Nonlinear effects in optical fibers: Origin, management and applications," *Progress in Electromagnetics Research, PIER*, vol. 73, 249–275, 2007.

23. N. Islam Khan, S. Hayder Choudhury, and A. Ahmed Roni, "A comparative study of the temperature depedence of lasing wavelength of conventional edge emitting strip laser and vertical cavity surface emitting laser," Proceedings of the International Conference on Data Communication Networking and International Conference on Optical Communication Systems, Athens, Greece, July 26–28, 2010.

24. J. Zhang, J. Yu, F. Li, X. Li, and Y. Wang, "Demonstration of single-carrier ETDM 400GE PAM-4 signals generation and detection," *IEEE Photonics Technology Letters*, vol. 27, no. 24, pp. 2543–2546, 2015.

25. SHF 603A MUX, Active digital module, SHF www.shf-communication.com/products/ high-speed-modules/

26. N. Eiselt, J. Wei, H. Griesser, A. Dochhan, M. Eiselt, J.-P. Elbers, J. J. V. Olmos, and I. T. Monroy, "First Real-Time 400G PAM-4 demonstration for inter-data center transmission over 100 km of SSMF at 1550 nm," Optical Fiber Communication, 2016, paper W1K.5.

27. Q. Hu, K. Schuh, M. Chagnon, F. Buchali, and H. Bülow, "84 GBd faster-than-Nyquist PAM-4 transmission using only linear equalizer at receiver," Optical Fiber Communication Conference (OFC), OSA Technical Digest, Optical Society of America, p. 2, 2019, paper W4I.

28. C. Yang, R. Hu, M. Luo, Q. Yang, C. Li, H. Li, S. Yu, "IM/DD-based 112-Gb/s/lambda PAM-4 transmission using 18-Gbps DML," IEEE Photonics Journal, vol. 8, no. 3, pp. 1–7, 2016.

29. S. Ummethala, J. N. Kemal, M. Lauermann, A. S. Alam, H. Zwickel, T. Harter, Y. Kutuvantavida, L. Hahn, S. H. Nandam, D. L. Elder, L. R. Dalton, W. Freude, S. Randel, and C. Koos, "Capacitively coupled silicon-organic hybrid modulator for 200 Gbit/s PAM-4 signaling," Conference on Lasers and Electro-Optics, OSA Technical Digest, Optical Society of America, 2019, paper JTh5B.2.

30. J. Zhang, J. S. Wey, J. Shi, J. Yu, Z. Tu, B. Yang, W. Yang, Y. Guo, X. Huang, and Z. Ma, "Experimental demonstration of unequally spaced PAM-4 signal to improve receiver sensitivity for 50-Gbps PON with power-dependent noise distribution," Optical Fiber Communication Conference, OSA Technical Digest, Optical Society of America, 2018, paper M2B.3.

31. J. Zhou, et al., "Transmission of 100-Gb/s DSB-DMT over 80-km SMF using 10-G class TTA and direct-detection," European Conference on Optical Communication, pp. 421–423, 2016.

32. Q. Hu, et al., "Advanced modulation formats for high performance short-reach optical interconnects," Optics Express, vol. 23, no. 3, pp. 3245–3259, 2015.

33. J. Zhang, J. Yu, J. Shi and H. Chien, "Digital dispersion pre-compensation and nonlinearity impairments pre- and post-processing for C-Band 400G PAM-4 transmission over SSMF based on direct-detection," European Conference on Optical Communication (ECOC), Gothenburg, pp. 1–3 2017.

34. J. Shi, J. Zhang, Y. Zhou, Y. Wang, N. Chi, and J. Yu, "Transmission performance comparison for 100-Gb/s PAM-4, CAP-16, and DFT-S OFDM with direct detection," Journal Lightwave Technology, vol. 35, pp. 5127–5133, 2017.

35. J. Shi, Y. Zhou, Y. Xu, J. Zhang, J. Yu, and N. Chi, "200-Gbps DFT-S OFDM using DD-MZM-based Twin-SSB with a MIMO-Volterra equalizer," IEEE Photonics Technology Letters, vol. 29, no. 14, pp. 1183–1186, 2017.

36. Zonglong He, Tianwai Bo, and Hoon Kim, "Probabilistically shaped coded modulation for IM/DD system," Optics Express, vol. 27, pp. 12126–12136, 2019.

37. A. Masuda, S. Yamamoto, H. Taniguchi, M. Nakamura, and Y. Kisaka, "255-Gbps PAM-8 transmission under 20-GHz bandwidth limitation using NL-MLSE based on Volterra filter," Optical Fiber Communication Conference (OFC), OSA Technical Digest, Optical Society of America, 2019, paper W4I.6.

38. N. Kikuchi, R. Hirai, and T. Fukui, "BER improvement of IM/DD higher-order optical PAM signal with precise non-linearity compensation," Optical Fiber Communication Conference (OFC), OSA Technical Digest, Optical Society of America, 2019, paper W4I.1.

39. J. Zhang, X. Li, Y. Xia, Y. Chen, X. Chen, J. Yu, and J. Xiao, "60-Gb/s CAP-64QAM transmission using DML with direct detection and digital equalization," Optical Fiber Communication, 2014, paper W1F.3.

40. J. Zhang, J. Yu, F. Li, N. Chi, Z. Dong, and X. Li, "11 × 5 × 9.3Gb/s WDM-CAP-PON based on optical single-side band multi-level multi-band carrier-less amplitude and phase modulation with direct detection," Optics Express, vol. 21, pp. 18842–18848, 2013.

41. T. Takahara, T. Tanaka, M. Nishiharra, Y. Kai, L. Li, Z. Tao, and J. C. Rasmussen, "Discrete Multi-Tone for 100 Gb/s optical access networks," Proceedings of Optical Fiber Communication, 2014, paper M2I.1.

42. D. P. Palomar and J. R. Fonollosa, "Practical algorithms for a family of waterfilling solutions," *IEEE Transactions on Signal Processing*, vol. 53, no. 2, pp. 686–695, 2005.

43. P. S. Chow, J. M. Cioffi, and J. A. C. Bingham, "A practical discrete multitone transceiver loading algorithm for data transmission over spectrally shaped channels," *IEEE Transactions on Communications*, vol. 43, no. 2, pp. 773–775, 1995.

44. A. Scaglione, S. Barbarossa, and G. B. Giannakis, "Optimal adaptive precoding for frequency-selective Nagakami-m fading channels," IEEE Vehicular Technology Conference, vol. 3, pp. 1291–1295, 2000.

45. B. Farhang-Boroujeny, "OFDM versus filter bank multicarrier," *IEEE Signal Processing Magazine*, vol. 28, no. 3, pp. 92–112, 2011.

46. V. Vakilian, T. Wild, F. Schaich, S. ten Brink, and J. Frigon, "Universal-filtered multicarrier technique for wireless systems beyond LTE," IEEE Globecom Workshops, pp. 223–228, 2013.

47. G. Fettweis, M. Krondorf, and S. Bittner, "GFDM – Generalized frequency division multiplexing," IEEE Vehicular Technology Conference, pp. 1–4, 2009.

48. A. Mecozzi, C. Antonelli, and M. Shtaif, "Kramers–Kronig coherent receiver," *Optica*, vol. 3, no. 11, pp. 1220–1227, 2016.

49. X. Chen, C. Antonelli, S. Chandrasekhar, G. Raybon, J. Sinsky, A. Mecozzi, M. Shtaif, and P. Winzer, "218-Gb/s single-wavelength, single-polarization, single-photodiode transmission over 125-km of standard singlemode fiber using Kramers-Kronig detection," Proceedings of Optical Fiber Communication, 2017, paper Th5B.6.

50. Z. Li, M. S. Erkılınç, S. Pachnicke, H. Griesser, R. Bouziane, B. C. Thomsen, P. Bayvel, and R. I. Killey, "Signal-signal beat interference cancellation in spectrally-efficient WDM direct-detection Nyquist-pulse-shaped 16-QAM subcarrier modulation," *Optics Express*, vol. 23, no. 18, pp. 23694–23709, 2015.

51. H. Shi, P. Yang, C. Ju, X. Chen, and J. Bei, "SSBI cancellation based on time diversity reception in SSB-DD-OOFDM transmission systems," Proceedings of Conference on Lasers & Electro-Optics, 2014, paper JTh2A.14.

52. J.-H. Yan, Y.-W. Chen, K.-H. Shen, and K.-M. Feng, "An experimental demonstration for carrier reused bidirectional PON system with adaptive modulation DDO-OFDM downstream and QPSK upstream signals," *Optics Express*, vol. 21, no. 23, pp. 28154–28166, 2013.

53. C. Antonelli, A. Mecozzi, M. Shtaif, X. Chen, S. Chandrasekhar, and P. J. Winzer, "Polarization multiplexing with the Kramers-Kronig receiver," *Journal of Lightwave Technology*, vol. 35, no. 24, pp. 5418–5424, 2017.

54. S. An, Q. Zhu, J. Li, and Y. Su, "Modified KK receiver with accurate field reconstruction at low CSPR condition," Proceedings of Optical Fiber Communication, 2019, paper M1H.3.

55. S. Fan, Q. Zhuge, Z. Xing, K. Zhang, T. M. Hoang, M. Morsy-Osman, M. Y. S. Sowailem, Y. Li, J. Wu, and D. V. Plant, "264 Gb/s twin-SSB-KK direct detection transmission enabled by MIMO processing," Proceedings of Optical Fiber Communication, 2018, paper W4E.5.

56. Y. Zhu, X. Ruan, Z. Chen, M. Jiang, K. Zou, C. Li, and F. Zhang, "4×200Gb/s Twin-SSB Nyquist subcarrier modulation WDM transmission over 160km SSMF with direct detection," Proceedings of Optical Fiber Communication, 2017, paper Tu3I.2.

57. J. Zhang, J. Shi, and J. Yu, "The best modulation format for 100G short-reach and metro networks: DMT, PAM-4, CAP, or duobinary?" SPIE Proceedings Volume 10560, Metro and Data Center Optical Networks and Short-Reach Links; 1056002, 2018.

3 Digital Coherent Optical Technologies

Zhensheng Jia and Mu Xu

CONTENTS

3.1 INTRODUCTION

Coherent optical receivers have initially received significant research interest in the 1980s [1–4]. During the period, no optical preamplification was used in front of the receiver, which results in detection that was mostly limited by the thermal noise of the photodiode and electrical amplifiers following square-law detection [5]. As the local oscillator (LO) signal typically has a much higher power than the received signal, it can be used for coherent amplification gain and can extract the phase information of the signal. Hence, for coherent detection, the receiver sensitivity is only limited by shot noise. This can potentially increase the receiver sensitivity with more than 20 dB in comparison to optically unamplified direct detection, and the linear electrical-to-optical field translation enables more independent degrees of freedom in modulation format selections for much higher spectral efficiency (SE).

However, the first commercial optical transmission systems were implemented using intensity-modulation and direct-detection (IM-DD). The invention of erbium-doped fiber amplifiers (EDFAs) made the shot-noise limited receiver sensitivity of the coherent receiver less significant [6]. This is because the optical signal-to-noise ratio (OSNR) transmitted through the amplifier chain is determined from the accumulated amplified spontaneous emission rather than the shot noise. In addition, even in unrepeated transmission systems, the EDFA used as a low-noise preamplifier eliminated some of the limitations of coherent detection that existed at that time, such as active polarization and carrier phase control [7]. Both the surge in EDFA-based optical communication solutions and the challenges with coherent detection interrupted research-and-development activities in coherent optical communications for nearly 20 years. Even though the increase in capacity enabled by EDFAs and dense wavelength division multiplexing (DWDM) [8–10] has scaled well in the past, a hard limit on capacity exists while non-coherent on–off keying (OOK) modulation is used, given that the maximum achievable information SE in theory is 1 bit/s/Hz. During the development of 40-Gb/s transmission systems, the need to maintain compatibility with the International Telecommunication Union (ITU) frequency grid of 50 GHz for WDM systems and reduce the need for extremely high bandwidth electronic components became apparent. Furthermore, the direct-detection system shows the limited ability to equalize the fiber impairments, such as polarization mode dispersion (PMD). Due to these pressures, coherent detection with advanced modulation formats, capable of transmitting more than 1 bit per symbol and effectively equalizing the fiber impairments, became a highly desirable technical advancement [11].

The development and maturity of high-speed digital application-specific integrated circuits (ASICs) have offered the possibility of using digital signal processing (DSP) capability to retrieve the linear in-phase and quadrature components of the complex amplitude of the optical carrier in a stable manner [12]. While an optical phase-locked loop (OPLL) that locks the LO phase to the signal phase is not needed, DSP circuits are becoming faster at an increasing rate and providing us with simple and efficient means for equalizing fiber-transmission impairments, demultiplexing two polarizations, and estimating the carrier phase. In short, the introduction and integration of DSP within coherent transceiver is the enabler for the rebirth of coherent optics. Employing DSP opens the door to all sorts of filtering and impairment compensation. Furthermore, only coherent detection permits convergence to the ultimate Shannon limits of SE in theory.

The first real-time coherent system was reported in 2008 for 40 Gb/s that used dual-polarization quadrature phase-shift-keying (DP-QPSK) modulation format [13]. Clock recovery, carrier recovery, polarization-and-PMD tracking and compensation, and chromatic dispersion compensation (CDC) are all done digitally, requiring 12 trillion (12×10^{12}) integer operations per second in the ASIC implementation [13]. Through the adoption of high-order modulation formats, higher SEs can be reached through the reduced symbol rate.

3.2 ADVANCED MODULATION FORMATS

As described earlier, the use of coherent detection and DSP enable optical system that fully leverage the benefits of advanced modulation formats and provide previously unavailable functionality in systems with direct detection, such as the use of phase and the polarization dimensions as independent orthogonal means to convey information [14]. The migration from traditionally used OOK modulation formats to formats with more bits per symbol and the ability to control the signal spectrum through pulse shaping lead to a reduction of the symbol rate and narrowed spectral widths. Therefore, higher SEs, per fiber capacities, and then cost per bit can be realized. This is the main motivation for a system upgrade to higher order modulation format and detection schemes. From a system perspective, there are multiple optimization parameters when considering the selection of the modulation formats, including the following:

- receiver power and/or OSNR sensitivity
- tolerance to fiber chromatic dispersion, nonlinearity, or PMD and polarization dependent loss (PDL)
- achievable SE
- resistance to filter narrowing impact
- complexity/cost of optical transmitter and receiver modules

The importance of each of these items varies widely between network types. Metro access systems generally emphasize the cost, complexity, and receiver sensitivity since the impact of fiber transmission impairments are relatively small compared to long-haul and ultra-long-haul optical systems [15].

It is possible to derive theoretical limits on the performance of different modulation formats based upon the use of optimal transmitters and receivers, and performance being purely impaired by additive white Gaussian noise (AWGN). The transmitter and receiver are assumed to have ideal matched filters, providing a signal that is band-limited at the Nyquist frequency [16]. The noise level is described here using the signal processing convention of E_b/N_0, where E_b is the mean energy per transmitted bit and N_0 is the mean noise energy per symbol. This metric enables comparison between modulation formats with differing parameters but at identical bit rates, as the SNR is normalized to the number of bits per symbol of the modulation format (unlike, e.g. E_s/N_0). If we convert this to the optical convention of OSNR over a bandwidth of 12.5 GHz (0.1 nm at 1,550 nm), the corresponding formula is obtained:

$$SNR = \frac{E_b}{N_0} = 2 * OSNR * \frac{12.5 \, \text{GHz}}{R_{line\_rate}} \tag{3.1}$$

where R_{line_rate} is the raw data rate of the optical link. As a result of an increasing number of bits per symbol, noise performance degrades as the Euclidean distances between the symbols become smaller. High-order quadrature amplitude modulation (QAM) formats exhibit a significantly better noise performance than high-order phase modulation formats for a certain number of bits per symbol [17–19], Square QAM formats in particular, due to the more optimum allocation of symbols on the complex plane. Square 16-QAM has an OSNR performance gain of about 4 dB compared to 16 PSK formats, for instance. Laser linewidth requirements increase with an increasing number of phase states [20], since a certain level of laser phase noise is more critical for closer phase distances. In addition, the reduction of the symbol rate makes the laser phase noise more critical for modulation formats with a higher number of bits per symbol. In systems with direct detection, the linewidth requirements are relatively relaxed.

As coherent optical communication systems have moved to 600 Gb/s and beyond interface rates [21–23], the pursuit of capacity-limit-approaching advanced modulation formats is under intensive investigation for the remaining marginal performance gains to close the gap toward the Shannon limit. Multi-dimensional-coded modulation is considered. It is the joint optimization of the coding and modulation schemes employed in a communication system for a given channel. It can show good transmission performance with the challenge of major forward error correction (FEC) redesign. Geometric [24] and probabilistic shapings [25,26] are the other two frontiers that have gained popularity in recent years for ultra-high capacity or ultra-long transmission distance optical systems.

- Geometric shaping involves multiple rings with equidistant points on each ring with uniform probability distribution across all points. Points on ring and distance between those are variable such that the overall distribution approaches the desired one. It can have a 0.3–0.7-dB practical gain with the challenge of significant DSP redesign.
- Probabilistic shaping has the constellation on a uniform grid with differing probabilities per constellation point. The fundamental principal is that there is a signal distribution that maximizes the achievable information rate for a given channel. The most commonly used channel for optical communications is the AWGN channel, which means an optimal signal distribution is also Gaussian. It has theoretically the best sensitivity that offers gains of up to 1.53 dB in SNR through this Gaussian-shaped constellations. The major challenge is the implementation complexity of the demapping process.

3.3   POLARIZATION-MULTIPLEXED (PM) QAM OPTICAL TRANSMITTER

There are many possible implementations for an optical polarization-multiplexed (PM)-QAM transmitter. In this section, the most commonly used transmitter architecture is presented and is compatible to arbitrary QAM constellation generations. The fundamental building blocks in the transmitter are tunable lasers and modulators. The tunable laser technologies include Distributed Feedback (DFB) Lasers, Sampled Grating Distributed Bragg Reflector (SG-DBR), and External Cavity Lasers (ECLs). The typical output power and linewidth of DFB and SG-DBR are 13 dBm and beyond 5 MHz, respectively. For coherent technology, both optical power and linewidth are critical factors for impacting signal performance. Higher power can compensate the insertion and modulation loss as the transmitter source and increase the coherent gain as the LO source. Meanwhile, narrower linewidth feature is especially important for coherent systems that use higher order modulation formats. Therefore, ECLs with wide tunable range (typically the whole C-band) are adopted in the initial long-haul and metro applications with an output power of 14–15 dBm and a linewidth from 30 to 500 kHz. For the optical access network, low-cost DFB or SG-DBR can be an option, which can provide a reasonable level of performance for QPSK-even 16-QAM signals to meet short-reach link budget requirement.

In terms of coherent optical modulator, currently, there are three materials for coherent IQ modulators, i.e. Lithium-Niobate ($LiNbO_3$), Indium-Phosphide (InP), and Silicon Photonics (SiPh). These types of commercial coherent IQ modulators are based on the configuration of parallel inner and outer traveling-wave Mach–Zehnder interferometers (MZI). LiNbO$_3$ modulators are based on the electro-optic effect, InP modulators mainly on the quantum-confined Stark effect, and silicon modulators on carrier depletion effect. Among these types of modulators, LiNbO$_3$ has been the most full-fledged technology. Its main drawback is that it requires a physically long MZI structure to achieve a reasonable V_π, which makes it hard to be implemented due to its smaller size than the other two. InP can fit inside a smaller package, but the challenge is that it requires a thermo-electric cooler to stabilize its operation and, therefore, has higher power consumption. SiPh-based modulators typically have high coupling loss that leads to the requirement of optical amplifier, such as additional EDFA. SiPh modulators, however, can allow wafer-level testing/probing and be hermetically sealed at wafer level, which may potentially enable a low-cost optical-module package.

As an example, Figure 3.1 presents an optical nested I/Q modulator with dual Mach–Zehnder modulators (MZMs) for QPSK and higher order QAMs modulation formats. The incoming light is equally split into two arms: the in-phase (I) and the quadrature (Q). In both paths, a field amplitude modulation is performed by operating the MZMs in the push–pull mode at the minimum transmission point. Moreover, a relative phase shift of $\pi/2$ is adjusted in one arm, for instance, by an additional phase modulator. This way, any constellation point can be reached in the complex IQ-plane after recombining the light of both branches.

As illustrated in Figure 3.1, the recombination of the two binary phase-shift-keying (BPSK) signals with $\pi/2$ phase difference yields a QPSK signal. In a similar way, as in QPSK generation, the mechanism for 16-QAM can be considered two PAM-4 signals inference with the phase shifted by $\pi/2$ (see Figure 3.1). It is a superposition of vectors on a complex plane. This BPSK or four-level PAM-4 signals can be obtained by digital-to-analog converters (DAC) in the commercial implementations,

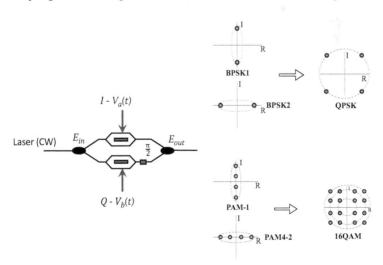

FIGURE 3.1 Architecture of Mach–Zehnder modulator and QPSK/16 QAM signal generation process.

or using a pulse-pattern generator in the lab. An arbitrary waveform generator is typically used in the proof-of-concept experimental verification. One of the most important parameters of the QAM signal modulation is the modulation loss, which depends on the following factors:

- Insertion loss of modulator
- Bias points of MZIs
- Driver swing and driver rise/fall times
- Modulation format
- Linearity of modulator
- Spectral shaping and pre-CDC

Back to the QPSK example, the transfer function of a single MZM is given by

$$E_{out} = E_{in}\cos(\pi V(t) / V_{\pi}) \tag{3.2}$$

where E_{in} is the input optical signal, E_{out} is the output optical signal of the MZM, $V(t)$ is the driving signal, and V_{π} is a characteristic of the MZM and determines the amplitude required for the driving signal. Thus, the transfer function of this IQ modulator is given by

$$E_{out} = E_{in}\sqrt{\cos^2(\pi V_a(t) / V_{\pi}) + \cos^2(\pi V_b(t) / V_{\pi})} * e^{jtan^{-1}\frac{\cos(\pi V_b(t)/V_{\pi})}{\cos(\pi V_a(t)/V_{\pi})}} \tag{3.3}$$

If $V_a(t)$ and $V_b(t)$ take on one of the two values $\{0, V_{\pi}\}$, the phase shift induced on the IQ modulator is one of the four values as shown in Table 3.1.

For generating dual-polarization modulation formats, typically two triple MZMs are used in parallel, each modulating an orthogonal polarization (see Figure 3.2). The two unmodulated carriers come from the same laser and are split into orthogonal linear polarizations with a polarization beam splitter (PBS), before the two independent polarization-modulated signals are multiplexed together with a polarization beam combiner. On the electrical signal flow, after the client to line rate and format-conversion process, the signals will be processed in transmitter DSP and DACs and then amplified by electrical drivers. These two pairs of amplified signals will drive the nested IQ optical modulator for coherent optical-signal generation.

TABLE 3.1

Phase Shifts in an IQ-QPSK Modulator

$V_a(t)$	$V_b(t)$	$\cos(\pi V_a(t) / V_{\pi})$	$\cos(\pi V_b(t) / V_{\pi})$	$\tan^{-1}\frac{\cos(\pi V_b(t)/V_{\pi})}{\cos(\pi V_a(t)/V_{\pi})}$
0	0	1	1	$\pi/4$
0	V_{π}	1	−1	$-\pi/4$
V_{π}	0	−1	1	$3\pi/4$
V_{π}	V_{π}	−1	−1	$5\pi/4$

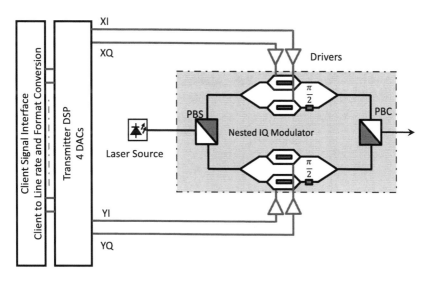

FIGURE 3.2   Optical transmitter for PM-QAM modulation formats.

3.4   COHERENT DETECTION SCHEMES

Homodyne/Intradyne/Heterodyne

In a coherent receiver, a LO is used to down-convert its sign with the incoming signal lightwave for demodulation. This coherent detection maps an entire optical field into the digital domain, therefore allowing detection of not only the amplitude of the signal but also its phase and the state of polarization (SOP). Depending on the intermediate frequency f_{IF} defined as $f_{IF} = f_s - f_{LO}$, coherent detection can be realized using a homodyne, intradyne, or heterodyne receiver as illustrated in Figure 3.3, where $Bandwidth_s$ is the optical signal bandwidth.

In a homodyne receiver [27], the LO and transmitter laser have the same frequency, and the phase difference should be zero (or a multiple of 2π). Homodyne detection is the ideal coherent-detection scheme in the sense that it allows for the optimal-receiver sensitivity with quantum noise limitation. Active control of the frequency and phase of the LO can be implemented in the optical domain via an optical OPLL. However, the OPLL operation puts stringent requirements on the laser linewidth for both the LO and transmitter laser, which makes homodyne detection difficult to realize with common laser sources.

Intradyne detection [28] is similar to homodyne detection, with the exception that frequency offset between the LO and transmitter laser exists, but the f_{IF} is chosen to fall within the signal band by roughly aligning the f_{LO} with f_s. Intradyne detection relies on the detection of both the in-phase and quadrature components of the received signal and is, therefore, also referred to as a phase-diversity receiver. Digital-phase-locking algorithms are needed to recover the modulation signal from its sampled I and Q components, which requires high-speed analog-to-digital conversion and DSP. As the signal has to be split into two components, intradyne detection

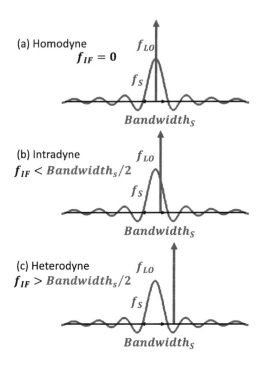

FIGURE 3.3 Three coherent detection schemes: (a) homodyne, (b) intradyne, and (c) heterodyne.

has a 3-dB sensitivity penalty in comparison to homodyne detection for shot-noise limited reception.

In heterodyne detection [29], the difference between the LO and transmitter laser frequency is higher than the electrical signal bandwidth and the entire optical signal spectrum is directly translated to an electrical bandpass signal centered at the f_{IF} for further electronic processing. Such a receiver has a reduced sensitivity of >3 dB in comparison to homodyne detection, as the signal energy with homodyne detection is twice the signal energy of a heterodyned signal. The linewidth requirements are, however, about an order of magnitude less stringent in comparison to homodyne detection, which makes the realization of such a receiver simple. The main drawback of heterodyne detection is that it requires a receiver bandwidth of at least twice the bit rate. A heterodyne receiver, therefore, needs broadband photodiodes and electrical amplifiers, which makes it challenging to realize high-speed transmission.

Thus, compared with heterodyne detection, intradyne detection has the advantage that the processing bandwidth is relaxed for the optical and electrical components. It is also noted that both heterodyne and intradyne detections that use single-ended detection are vulnerable to RIN from the LO if extra DSP is not implemented. This can be solved through balanced detection [30] that requires a total of four photodiodes and an optical hybrid with four outputs, each shifted by 90°.

3.5   COHERENT RECEIVER ARCHITECTURE

The fundamental concept behind coherent detection is to take the beating product of electric fields of the modulated signal light and the continuous-wave LO. To detect both the I/Q components of the signal light, a 90° optical hybrid is utilized. A key building block of such a hybrid is a 2 × 2 optical coupler with its property of a 90° phase shift between its direct-pass and cross-coupling outputs via multimode interference coupler. The combination of such optical couplers into the configuration shown in Figure 3.4, together with an additional 90° phase shift in one arm, can achieve a detection of real and imaginary parts. Balanced detection is usually introduced into the coherent receiver as a mean to suppress the DC component and maximize the signal photocurrent.

Output photocurrents from balanced photodetectors are then given as

$$I_I(t) = I_1(t) - I_2(t) = R\sqrt{P_s P_{LO}}\cos\{\varphi_s(t) - \theta_{LO}(t)\} \tag{3.4}$$

$$I_Q(t) = I_3(t) - I_4(t) = R\sqrt{P_s P_{LO}}\sin\{\varphi_s(t) - \theta_{LO}(t)\} \tag{3.5}$$

where R is the responsitivity of the photodiode, P_s and P_{LO} are the power of the optical fields for incoming and LO signal, respectively. The receiver thus leads to the recovery of both the sine and cosine components. It is possible to estimate the phase noise $\theta_{LO}(t)$ varying with time and restore the phase information $\varphi_s(t)$ through the following DSP on the intradyne-detected signal.

The schematic diagram of a polarization multiplexed coherent receiver is shown in Figure 3.5. Both the incoming PM signal and LO are split into two orthogonal polarizations using a PBS, after which the co-polarized signal and the local oscillator are mixed into two 90° optical hybrids to produce in-phase and quadrature components for each polarization. The four signals are then digitized by four analog-to-digital converters (ADCs) after which DSP can be performed for signal demodulation.

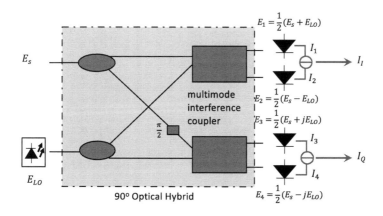

FIGURE 3.4   Configuration of phase-diversity coherent receiver.

Polarization-diversity
90° optical hybrid

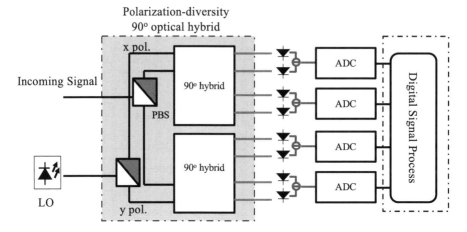

FIGURE 3.5   Configuration of phase- and polarization-diversity coherent receiver architecture.

3.6   DIGITAL EQUALIZATION ALGORITHMS

Current coherent optical transceivers now utilize DSP with the transmitter being responsible for modulation, pulse shaping and pre-equalization, and the receiver responsible for equalization, synchronization, and demodulation.

At the transmitter, the DSP, in conjunction with the DACs and FEC, converts the incoming data bits into a set of analog signals. As shown in Figure 3.6 in detail, transmitter DSP functions include symbol mapping and signal timing deskew adjustment, optional pre-distortion for dispersion or self-phase modulation, and software-programmable capability of supporting multiple modulation formats and encoding schemes. Transmitter DSP also allows compensating nonlinearities induced by the electrical driver and the optical modulator.

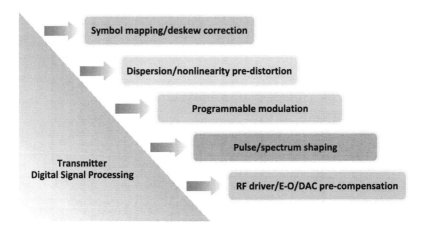

FIGURE 3.6   Transmitter-side DSP functions.

The major benefit of DSP in the transmitter is to perform pulse shaping and thus engineer the spectrum of signal, such as generating Nyquist pulse. Pulse shaping controls the spectrum to increase the SE or reduce the nonlinear impairments. Basically, there are three types of pulse shaping filters, sinc, raised-cosine, and Gaussian filters. Table 3.2 summarizes their corresponding transfer function, spectral and pulse shapes, and eye diagrams of QPSK-filtered driving signals [31].

A sinc-shaped filter is an ideal filter in the sense that its rectangular spectrum requires the minimum Nyquist channel bandwidth without inter-symbol interference (ISI). However, it is impractical because of its infinite time expansion. A raised-cosine filter and its associated root-raised-cosine filter (matched filter version) smoothly approach the frequency stop band and are practical implementations of ISI-free spectral prefiltering with a roll-off factor for tradeoff between the number of taps of impulse response and bandwidth occupancy. This raised-cosine filter is basically adopted to reduce channel crosstalk and/or increase the tolerance toward narrow filtering effect with limited achievable SE and without the need of post-compensation. A Gaussian filter, which is commonly used, has a smooth transfer function and no-zero crossings as shown in Table 3.1. One of the most attractive features of Gaussian filter is to significantly reduce the side lobes of signal spectrum and the availability of real products. To further increase the SE, this filter is used to perform spectral prefiltering to lead to super-Nyquist situation, in which the signal occupancy is smaller than channel spacing. In the meantime, the introduced severe ISI and/or the inter-channel interference (ICI) can be post-compensated with effective algorithms.

Transmitter-side DSP enables more flexibility not only in channel impairment pre-compensation but also in software configuration for elastic optical networks. DAC are then used to generate multi-level analog electrical drive signals for optical modulators. In correspondence to the operation of the transmitter, the major advantage of receiver-side DSP stems from the ability to arbitrarily manipulate the electrical field after the ADC enables the sampling of the signal into digital domain. As shown in Figure 3.7, the fundamental DSP functionality in a digital coherent receiver for PM-QAM signals can be illustrated by the following flow of steps and their correlations from structural and algorithmic level of details.

First, the four digitized signals (i.e. in-phase and quadrature components for each polarization) after an ADCs are passed through the block for the compensation of front-end imperfections. The imperfections may include timing skew between the four channels due to the difference in both optical and electrical path lengths within a coherent receiver. Other types of front-end imperfections can be the difference between the output powers of four channels due to different responses of positive–intrinsic–negative (PIN) and transimpedance amplifiers (TIAs) in the receiver, and quadrature imbalance because the optical hybrid may not exactly introduce a 90° phase shift.

Second, the major channel-transmission impairments are compensated through power-efficient digital filters, in particular, chromatic dispersion and PMD. Based on different time scales of the dynamics of these impairments, the static equalization for CDC is performed first because of its independent SOP and modulation format and the impact on the subsequent blocks before the chromatic dispersion estimation is needed to achieve accurate compensation. Then the clock recovery for symbol synchronization can be processed to track the timing information of incoming

TABLE 3.2
Three Pulse Shaping Filters with Different Transfer Functions

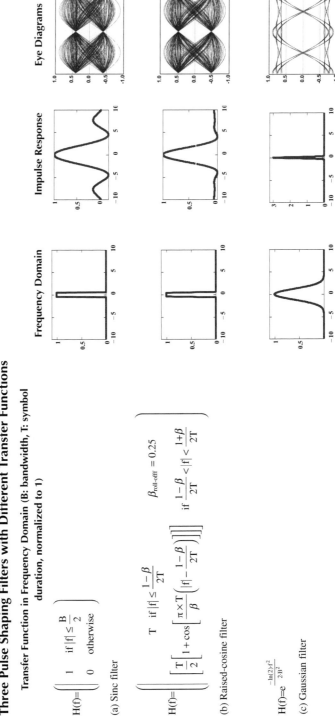

Transfer Function in Frequency Domain (B: bandwidth, T: symbol duration, normalized to 1)

$$H(f) = \begin{cases} 1 & \text{if } |f| \leq \dfrac{B}{2} \\ 0 & \text{otherwise} \end{cases}$$

(a) Sinc filter

$$H(f) = \begin{cases} T & \text{if } |f| \leq \dfrac{1-\beta}{2T} \\ \left[\dfrac{T}{2}\left[1+\cos\left[\dfrac{\pi \times T}{\beta}\left(|f|-\left(\dfrac{1-\beta}{2T}\right)\right)\right]\right]\right] & \text{if } \dfrac{1-\beta}{2T} < |f| < \dfrac{1+\beta}{2T} \end{cases}$$

$$\beta_{\text{roll-off}} = 0.25$$

(b) Raised-cosine filter

$$H(f) = e^{\frac{-\ln(2) \cdot f^2}{2 \cdot B^2}}$$

(c) Gaussian filter

Frequency Domain   Impulse Response   Eye Diagrams

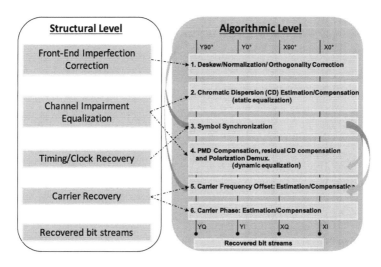

FIGURE 3.7 DSP flow in a digital optical coherent receiver.

samples. Note that it is possible to perform joint processing between the blocks of clock recovery and polarization demultiplexing for achieving the symbol synchronization after equalizing all channel impairments (see arrows in Figure 3.7). A fast-adaptive equalization is carried out jointly for two polarizations through a butterfly structure and the stochastic gradient algorithms. Then the frequency offset between the source laser and the LO is estimated and removed to prevent the constellation rotation at the intradyne frequency.

Finally, the carrier phase noise is estimated and removed from the modulated signal, which is then followed by symbol estimation and hard or soft-decision FEC for channel decoding. Note that for a particular digital coherent receiver, the ordering of DSP flow may differ slightly from those detailed in Figure 3.7 because of different design choices. Besides the feed-forward process, it is possible to incorporate feedback and perform joint processing among different process blocks such as clock recovery and polarization demultiplexing as mentioned above. It is also possible to perform the same functions by either training sequence-based data-aided or totally blinded algorithms.

The constellation evolutions show examples (Figure 3.8) of the received signal after linear transmission over uncompensated single-mode fiber (SMF) link with EDFAs only. Note that the results in this proposal are based on 32-GBaud rate and that the impairments of frequency offset of 0.1 GHz, 100-kHz linewidth of transmitter laser, and LO are induced with 20,000 ps/nm accumulated chromatic dispersion.

Coherent detection and DSP were the key enabling technologies in the development of 100G optical-transmission systems. The next-gen coherent optical systems will continue this trend with DSP playing even more ubiquitous role at both transmitter and receiver. The generic functions at a structural or function abstraction level are similar for all major commercial products. Nevertheless, at a deeper implementation level, different specific algorithms for each process block, used as various realizations, can address the same process block function.

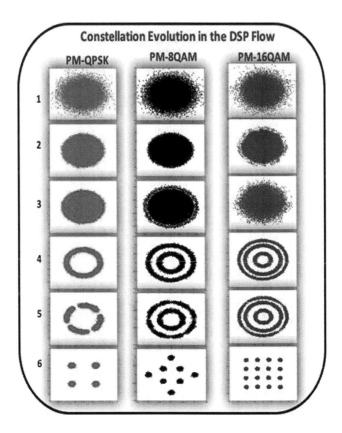

FIGURE 3.8   Constellation evolutions for QPSK/8 QAM/16 QAM signals.

3.7   APPLICATION-SPECIFIC INTEGRATED CIRCUIT

One of the most important building blocks in coherent optical transceiver is an ASIC for the DSP implementation [32,33]. The ASIC combines data converters between analog and digital domains and DSP for all necessary algorithmic-processing steps. In some cases, serializers/deserializers (Ser/Des) for client and line-rate-and-format conversion are also incorporated in a single ASIC device.

Data converters in the ASIC, i.e. DACs and ADCs, serve as the interface between the optical front-ends and DSP. The ultra-high speed and low-power ADC and DAC circuits are by far the most critical and challenging components, since their capabilities basically dictate how the overall optical transceiver can be implemented. DSP, in both transmitter and receiver, contains all the necessary steps to perform the desired signal generation and recovery in digital domain, including impairment equalization, clock and carrier recovery, and error-correction encoding and decoding. The coherent optical ASICs have been evolved in several generations across the optical industry:

1. First generation: The first DP-QPSK 100G coherent transceiver is based on 65-nm complementary metal–oxide–semiconductor (CMOS) process

nodes. The ADCs/DACs have a sampling rate ≥ 56 GSa/s, an effective number of bits (ENOB) ≥ 5.5 bits, and an analog bandwidth ≥ 16 GHz for ≥ 28 GBaud signal generation and detection. The ASIC contains more than 50 million (M) transistors.

2. Second generation: The single-channel data rate increases to 200G with 40-nm CMOS process nodes. The ADCs/DACs have a sampling rate 55–65 GSa/s, an ENOB ≥ 5.7 bits, and an analog bandwidth ≥ 19 GHz. The ASIC contains >70M transistors. Meanwhile, supporting multiple modulation formats with software programmable ability is made possible to trade off the capacity or SE with transmission distance.

3. Third generation: The channel data rate increases to 400G with 28-nm CMOS process nodes. The ADCs/DACs have a sampling rate 55–92 GSa/s, an ENOB ≥ 6 bits, and an analog bandwidth ≥ 26 GHz. The ASIC contains >200M transistors. From this generation, the optical industry has adopted two separate design paths for high-performance and low-power coherent ASICs. The first low-power ASIC was specially optimized to meet a power target compatible with a C-form factor pluggable (CFP) standard. The use of 1/2 and 1/4 rate modes is also adopted for different operation modes.

4. Fourth generation: The channel data rate increases to 600G with 16/14-nm CMOS process nodes. The semiconductor industry has migrated from standard bulk CMOS process to Fin-FET process. The ADCs/DACs have a sampling rate 34–128 GSa/s, an ENOB 5.5–6.5 bits, and an analog bandwidth ≥ 35 GHz. The ASIC contains >400M transistors.

5. Fifth generation: The channel data rate increases to 800G-1T with 7-nm CMOS process nodes. The ADCs/DACs have a sampling rate up to 140 GSa/s, an ENOB up to 8.5 bits, and an analog bandwidth ≥ 42 GHz. The ASIC targets >1,000M transistors.

Designing high-speed ADC/DAC is getting more complex from generation to generation using smaller geometries and ADC/DAC co-integration with the DSP and Ser/Des having a mixed signal design is making implementation even more challenging.

3.8   COHERENT TRANSCEIVER PLUGGABLE MODULE EVOLUTION

First coherent 100-Gb/s interfaces were built using a discrete line card architecture by large network-equipment manufacturers. Next to custom line cards, 100G was also defined in a standard line-interface module with a total power of less than 80 W in $5'' \times 7''$ Optical Internetworking Forum (OIF) compliant Multi-Source Agreement (MSA) modules for PM QPSK modulation signals. The intention of the 100-G MSA module is to support long-haul applications beyond 2,000 km. The second non-pluggable MSA module is $4'' \times 5''$, consuming power less than 40 W as shown in Figure 3.9. To provide high-density and low-cost coherent transceivers, hot pluggable modules are being implemented with small form factors. The power consumption for a coherent transceiver mainly includes the power used up by the ASICs, lasers, modulator drivers, and TIAs. In order to pack all the components in a small form factor, power dissipation needs to be reduced significantly.

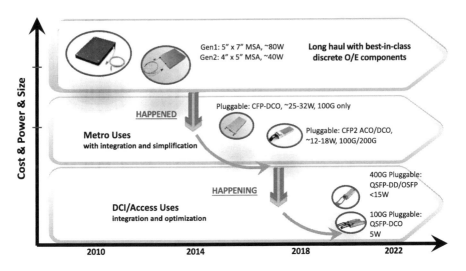

FIGURE 3.9   Evolution of digital coherent optical module.

There are two classes of pluggable transceivers for the optical transmission nodes [34]: one is the line-side transceivers for interface with long-haul or metro transmission optics and the other is the client-side transceivers to provide high-speed interconnects to the tributary ports of the network node. Today, the development of most of the client-side transceivers is focusing on CFP2, CSP, and Quad Small Form-Factor Pluggable (QSFP)+ formats without coherent optics. For the line-side transceiver with coherent optics, the CFP modules emerging initially are available and moving to CFP2 and CFP4 pluggable formats. The power requirement at client side for a CFP is <28–32 W, while a CFP2 has to fulfill a maximum of <9–12 W. Due to the strict power budget, in the line-side application, small form factor modules like the CFP2 can also be implemented with analog–host interfaces and the DSP–ASIC on an external board (CFP2-Analog Coherent Optics), and finally by the digital CFP2 (CFP2-Digital Coherent Optics (DCO)), which, similar to CFP, includes all optics and DSP/ASIC, but in half the size of the CFP. Photonic integration circuit appears to be a must for CFPx modules instead of separate LiNbO$_3$ modulators and InP receivers. To achieve the cost targets, the use of electronic integration and photonic integration needs to be implemented in order to reduce component count and improve manufacturability. Another dominant cost for the DSP and optics is the packaging; one can further reduce cost, power, and footprint by co-packaging the DSP and optics [35].

For the next-gen pluggable module, there are two major form factors being implemented for 400G transceivers from the largest to smallest: QSFP Double Density (QSFP-DD) and Octal Small Formfactor Pluggable (OSFP). QSFP-DD is a new module similar to current QSFP, but with an additional row of contacts providing an eight-lane electrical interface. The term "Double Density" refers to the doubling of the number of high-speed electrical interfaces compared to the regular QSFP28 module. Therefore, QSFP-DD expands on the QSFP pluggable form factor, a widely adopted four-lane electrical interface. One of the major benefits of QSFP-DD is backward compatibility with QSFP+ (40G), QSFP28 (100G), and QSFP56 (200G). Backward compatibility is critically important to the industry. Since ASICs are designed to support multiple interface rates, it is critically important that the system can take advantage of this. However, OSFP allows more power (12–15 W) than the QSFP-DD (7–12 W) because of larger size and direct integration of thermal management. For the 100G-only next-gen pluggable module in access application, QSFP28-DCO is one favorable option, which is compatible with most current 100G client interfaces. The target power dissipation is less than 5 W for I-temp application.

Coherent optical technology has been well established as the dominant technology of choice in long haul for a decade ago and for the last 4 years in metro. Recent technology advances and ongoing price drop further open the window of opportunity for the application of coherent optics in access networks. It is envisioned that the migration of coherent optics from long haul and metro to access domain will require optimization and deeper integration to decrease the cost and complexity further.

3.9 COHERENT OFDM

Except single-carrier modulation formats, the multi-carrier ones like orthogonal frequency division multiplexing (OFDM) can also be used in the coherent system. As introduced in a previous section, the generation of single carrier-based QAM signal in a coherent system is simple and straightforward, but the DSP at the receiver is quite sophisticated. Those DSP procedures, including carrier frequency offset (CFO) estimation, phase noise compensation, and channel equalization, are accomplished based on blind algorithms, meaning that the algorithms are executed on the basis of the statistical property of the signal without knowing the specific information it carries. However, there are some drawbacks of single-carrier modulation worth mentioning. First, clock recovery suffers from severe skew and timing offset among multiple data streams in coherent optical systems. Second, DSP complexity is high especially for CFO and carrier phase-noise estimation. If high-order modulation formats beyond 16-ary QAM are used, the complexity of blind DSP increases even further.

However, as a multi-carrier modulation format, OFDM is distinguished by its higher SE and flexibility, e.g. to load different modulation formats and power levels at different subcarriers to maximize the total capacity or minimize the overall power consumption. When combining with pilots or training symbols, the DSP for eliminating CFO and phase noise becomes simpler and faster. Signal recovery for high-order QAM becomes more feasible because, in most cases, simple formats, like QPSK, are modulated onto the pilots, and the channel information extracted from them can be directly applied to the subcarriers carrying more complex modulation formats. The multi-carrier feature empowers OFDM to be adaptive toward the different channel conditions and to work better under the influence of fading of strong-frequency domains, as it can leave some unloaded subcarriers when the channel response is degraded at certain frequency points. Nevertheless, one of the major drawbacks of OFDM lies in its high peak-to-average-power ratio (PAPR), which requires a high-power, low-noise electrical amplifier with a large dynamic range to drive the electro-optical modulator; otherwise, the peak and bottom of the waveform may be clipped, which introduces signal distortion. The strict requirements on the amplifier seriously reduce its power efficiency. The issue could be mitigated through some DSP techniques, such as discrete-Fourier-transform-spread OFDM [36], but the PAPR is still much worse than single-carrier modulation formats.

The DSP of coherent optical OFDM is similar to that of DMT discussed in Chapter 2. There are also several differences between them. In first place, there are four streams of signals in coherent optical OFDM with independent in-phase (I) and quadrature (Q) phases as well as X and Y polarizations. Thus, in order to correctly recover the signals, the synchronization of timing and clock recovery is very critical to guarantee the alignment of the four streams of data and to make sure that they are processed with the same clock. Thanks to the cyclic prefix (CP) of OFDM, a small-time mismatch can be absorbed by the CP if it is within its protection window. By allocating a part of the subcarriers as pilot tones, the phase and CFO can be effectively estimated and rectified. The pilot tone–based carrier recovery algorithms have been extensively investigated [13,37–39]. In [40,41], multiple pilot subcarriers are distributed in each OFDM symbol. By comparing the received and the original signals carried by each subcarrier, the phase rotations can be estimated. After taking the average across all the pilot subcarriers, phase-estimation accuracy can be improved. However, the above algorithm assumes that the phase drift within one OFDM symbol can be considered a constant and common to all the subcarriers. However, two problems are hiding behind such an assumption: first, except the common phase error, there are phase distortions contributed by the interference from the neighboring subcarriers. Intercarrier interference estimation and interpolations to mitigate this interference is applied [38]. Second, there may be phase variance within one OFDM symbol duration, which is especially significant at slower baud rates. In order to deal with such intra-symbol fast-phase fluctuations, pilot tone surrounded by unloaded subcarriers as band gaps is proposed [39]. As performance is improved, computational complexity increases and SE is also sacrificed. The intercarrier interference, however, is still not thoroughly eliminated. In general, compared with single-carrier modulation format, where every QAM symbol can be used for carrier recovery using blind methods, the performance of coherent optical OFDM is limited especially when suffering from fast phase fluctuations while using a non-ideal optical local oscillator. The high PAPR also weakens the system's resistance against nonlinear distortions from both electrical amplification and fiber nonlinearities. High PAPR also increases the cost in hardware because of the need for high-resolution DAC and ADC. These practical implementation disadvantages dilute the benefits brought by OFDM. In most of today's high-speed optical access networks, OFDM cannot be considered the first choice. A number of challenges faced by coherent optical OFDM systems are yet to be elucidated.

REFERENCES

1. B. Glance, "Polarization independent coherent optical receiver," *Journal of Lightwave Technology*, vol. 5, no. 2, pp. 274–276, 1987.
2. L. G. Kazovsky, "Phase- and polarization-diversity coherent optical techniques," *Journal of Lightwave Technology*, vol. 7, no. 2, pp. 279–292, 1989.
3. R. A. Linke and A. H. Gnauck, "High-capacity coherent lightwave systems," *Journal of Lightwave Technology*, vol. 6, no. 11, pp. 1750–1769, 1988.

4. A. W. Davis, S. Wright, M. J. Pettitt, J. P. King, and K. Richards, "Coherent optical receiver for 680 Mbit/s using phase diversity," *Electronics Letters*, vol. 22, no. 1, pp. 9–11, 1986.

5. J. Bowers and C. Burrus, "Ultrawide-band long-wavelength p-i-n photodetectors," *Journal of Lightwave Technology*, vol. 5, no. 10, pp. 1339–1350, 1987.

6. W. J. Miniscalco, "Erbium-doped glasses for fiber amplifiers at 1500 nm," *Journal of Lightwave Technology*, vol. 9, no. 2, pp. 234–250, 1991.

7. I. Stanley, G. Hill, and D. Smith, "The application of coherent optical techniques to wideband networks," *Journal of Lightwave Technology*, vol. 5, no. 4, pp. 439–451, 1987.

8. B. Mukherjee, "WDM optical communication networks: progress and challenges," *IEEE Journal on Selected Areas in Communications*, vol. 18, no. 10, pp. 1810–1824, 2000.

9. Y. Sun, A. K. Srivastava, J. Zhou, and J. W. Sulhoff, "Optical fiber amplifiers for WDM optical networks," *Bell Labs Technical Journal*, vol. 4, pp. 187–206, 1999.

10. A. Banerjee, Y. Park, F. Clarke, H. Song, S. Yang, G. Kramer, K. Kim, and B. Mukherjee, "Wavelength-division-multiplexed passive optical network (WDM-PON) technologies for broadband access: a review," *Journal of Optical Networking*, vol. 4, no. 11, pp. 737–758, 2005.

11. K. Kikuchi, "Fundamentals of coherent optical fiber communications," *Journal of Lightwave Technology*, vol. 34, no. 1, pp. 5115–0111, 2016.

12. C. Laperle and M. O'Sullivan, "Advances in high-speed DACs, ADCs, and DSP for optical coherent transceivers," *Journal of Lightwave Technology*, vol. 32, no. 4, pp. 629–643, 2014.

13. H. Sun, K.-T. Wu, and K. Roberts, "Real-time measurements of a 40 Gb/s coherent system," *Optics* Express, vol. 16, no. 2, pp. 873–879, 2008.

14. P. J. Winzer, "High-spectral-efficiency optical modulation formats," *Journal of Lightwave Technology*, vol. 30, no. 24, pp. 3824–3835, 2012.

15. Z. Jia, L. A. Campos, C. Stengrim, J. Wang, C. Knittle, "Digital coherent transmission for next-generation cable operators' optical access networks," Oct. SCTE/ISBE Cable-Tec Expo'17, 2017.

16. G. Bosco, A. Carena, V. Curri, P. Poggiolini, and F. Forghieri, "Performance limits of Nyquist-WDM and CO-OFDM in high-speed PM-QPSK systems," *IEEE Photonics Technology Letters*, vol. 22, no. 15, pp. 1129–1131, 2010.

17. X. Zhou, "An improved feed-forward carrier recovery algorithm for coherent receivers with M-QAM modulation format," *IEEE Photonics Technology Letters*, vol. 22, no. 14, pp. 1051–1053, 2010.

18. M. Seimetz, "Laser linewidth limitations for optical systems with high-order modulation employing feed forward digital carrier phase estimation," Proceedings of Optical Fiber Communications, 2008, paper OTuM2.

19. K. Kikuchi, "Fundamentals of coherent optical fiber communications," *Journal of Lightwave Technology*, vol. 34, no. 1, pp. 157–179, 2016.

20. I. Fatadin, D. Ives and S. J. Savory, "Laser linewidth tolerance for 16-QAM coherent optical systems using QPSK partitioning," *IEEE Photonics Technology Letters*, vol. 22, no. 9, pp. 631–633, 2010.

21. M. Xu, J. Zhang, H. Zhang, Z. Jia, J. Wang, L. Cheng, A. Campos, and C. Knittle, "Multi-stage machine learning enhanced DSP for DP-64QAM coherent optical transmission systems," Proceedings of European Conference on Optical Communication, 2019, paper M2H.1.

22. Y. R. Zhou, K. Smith, R. Payne, A. Lord, L. Raddatz, T. V. De Velde, C. Colombo, E. Korkmaz, M. Fontana, and S. Evans, "1.4 Tb real-time alien superchannel transport demonstration Over 410 km installed fiber link using software reconfigurable DP-16 QAM/QPSK," *Journal of Lightwave Technology*, vol. 33, no. 3, pp. 639–644, 2015.

23. H.-C. Chien, J. Yu, Y. Cai, B. Zhu, X. Xiao, Y. Xia, X. Wei, T. Wang, and Y. Chen, "Approaching terabits per carrier metro-regional transmission using beyond-100GBd coherent optics with probabilistically shaped DP-64QAM modulation," *Journal of Lightwave Technology*, vol. 37, no. 8, pp. 1751–1755, 2019.

24. A. I. Abd El-Rahman and J. C. Cartledge, "Multidimensional geometric shaping for QAM constellations," Proceedings of European Conference on Optical Communication, 2017, paper 8346118.

25. F. Buchali, F. Steiner, G. Böcherer, L. Schmalen, P. Schulte and W. Idler, "Rate adaptation and reach increase by probabilistically shaped 64-QAM: an experimental demonstration," *Journal of Lightwave Technology*, vol. 34, no. 7, pp. 1599–1609, 2016.

26. F. Buchali, G. Böcherer, W. Idler, L. Schmalen, P. Schulte and F. Steiner, "Experimental demonstration of capacity increase and rate-adaptation by probabilistically shaped 64-QAM," Proceedings of European Conference on Optical Communication, pp. 1–3, 2015.

27. E. Ip, A. P. T. Lau, D. J. F. Barros, and J. M. Kahn, "Coherent detection in optical fiber systems," *Optics Express*, vol. 16, no. 2, pp. 753–791, 2008.

28. F. Derr, "Coherent optical QPSK intradyne system: concept and digital receiver realization," *Journal of Lightwave Technology*, vol. 10, no. 9, pp. 1290–1296, 1992.

29. S. L. Jansen, I. Morita, T. C. W. Schenk, N. Takeda and H. Tanaka, "Coherent optical 25.8-Gb/s OFDM transmission over 4160-km SSMF," *Journal of Lightwave Technology*, vol. 26, no. 1, pp. 6–15, 2008.

30. Y. Painchaud, M. Poulin, M. Morin, and M. Têtu, "Performance of balanced detection in a coherent receiver," *Optics Express*, vol. 17, no. 5, pp. 3659–3672, 2009.

31. Z. Jia, Y. Cai, H.-C. Chien, and J. Yu, "Performance comparison of spectrum-narrowing equalizations with maximum likelihood sequence estimation and soft-decision output," *Optics Express*, vol. 22, no. 5, pp. 6047–6059, 2014.

32. W. Forysiak and D. S. Govan, "Progress toward 100-G digital coherent pluggable using InP-based photonics," *Journal of Lightwave Technology*, vol. 32, no. 16, pp. 2925–2934, 2014.

33. Z. Zhang, C. Li, J. Chen, T. Ding, Y. Wang, H. Xiang, Z. Xiao, L. Li, M. Si, and X. Cui, "Coherent transceiver operating at 61-Gbaud/s," *Optics Express*, vol. 23, no. 15, pp. 18988–18995, 2015.

34. T. Duthel, P. Hermann, T. W. von Mohrenfels, J. Whiteaway, and T. Kupfer, "Challenges with pluggable optical modules for coherent optical communication systems," Proceedings of Optical Fiber Communications, 2014, paper W3K.2.

35. C. R. Doerr, L. L. Buhl, Y. Baeyens, R. Aroca, S. Chandrasekhar, X. Liu, L. Chen, and Y.-K. Chen, "Packaged monolithic silicon 112-Gb/s coherent receiver," *IEEE Photonics Technology Letters*, vol. 23, no. 12, pp. 762–764, 2011.

36. K. Kikuchi, "Fundamentals of coherent optical fiber communications," *Journal of Lightwave Technology*, vol. 34, no. 1, pp. 157–179, 2016.

37. I. C. Wong, O. Oteri, and W. Mccoy, "Optimal resource allocation in uplink SC-FDMA systems," *IEEE Transactions on Wireless Communications*, vol. 8, no. 5, pp. 2161–2165, 2009.

38. T. M. Schmidl and D. C. Cox, "Robust frequency and timing synchronization for OFDM," *IEEE Transactions on Communications*, vol. 45, no. 12, pp. 1613–1621, 1997.

39. X. Yi, W. Shieh and Y. Tang, "Phase estimation for coherent optical OFDM," *IEEE Photonics Technology Letters*, vol. 19, no. 12, pp. 919–921, 2007.

40. V. Syrjala, M. Valkama, N. N. Tchamov, and J. Rinne, "Phase noise modelling and mitigation techniques in OFDM communications systems," 2009 Wireless Telecommunications Symposium, Prague, pp. 1–7, 2009.

41. S. Randel, S. Adhikari, and S. L. Jansen, "Analysis of RF-pilot-based phase noise compensation for coherent optical OFDM systems," *IEEE Photonics Technology Letters*, vol. 22, no. 17, pp. 1288–1290, 2010.

4 Coherent Technology Transition to Access Networks

Zhensheng Jia and Lin Cheng

CONTENTS

4.1 INTRODUCTION

Coherent optics initially received significant research interest in the 1980s because of its high receiver sensitivity through coherent amplification by a local oscillator, but its use in commercial systems has been hindered by the additional complexity of active phase and polarization tracking. In the meantime, the emergence of a cost-effective erbium-doped fiber amplifier (EDFA) as an optical pre-amplifier reduced the urgency to commercialize coherent detection, because EDFAs and wavelength-division multiplexing (WDM) extended the reach and capacity, respectively. Traffic demand, combined with the requirement to reduce cost per bit per Hz, or spectral efficiency (SE) increases, as well as advancements in complementary metal–oxide–semiconductor (CMOS)-processing nodes and powerful digital signal processing (DSP), led to the renaissance of coherent optics technology back in the year of 2008 [1].

Commercial coherent optical technology was first deployed in long-haul applications in 2010 to overcome fiber impairments that required complex compensation techniques when using direct-detection receivers. The first-generation coherent optical systems were based on a single-carrier polarization division multiplexed quadrature phase-shift-keying (QPSK) modulation format, and the achieved SE is 2 bit/s/Hz over conventional 50-GHz optical grid; since then the system capacity has been increased to around 10 Tb/s in the C-band transmission window [2]. By leveraging further development of CMOS processing, reduction in design complexity, and price decreases on opto-electro components, coherent solutions are moving from long haul and metro to access networks. This migration model has been experienced by the optical industry before: the dense (DWDM) system technology started in the long haul and then migrated to metro and edge access; forward error correction (FEC) encoding and decoding followed the same pattern. Benefiting from initial long-haul technology development, coherent optics for access networks will be the next natural progression [3,4].

4.2   COHERENT OPTICS DRIVERS
FOR ACCESS NETWORKS

Coherent detection for access networks enables superior receiver sensitivity that allows extended power budget and high-frequency selectivity for closely spaced DWDM/ultra-DWDM channels without the need of narrow-band optical filters. Moreover, the multi-dimensional signal recovered by coherent detection provides additional benefits to compensate linear transmission impairments such as chromatic dispersion (CD) and polarization mode dispersion (PMD) and efficiently utilizes the spectral resource, benefiting future network upgrades through the use of multi-level advanced modulation formats. However, there are several engineering challenges of introducing digital coherent technologies into access networks. The coherent technology in long-haul optical system utilizes best-in-class discrete photonic and electronic components, such as state-of-the-art digital-to-analog converter (DAC)/ analog-to-digital converter (ADC) and DSP application-specific integrated circuit (ASIC) based on the most recent CMOS process. The coherent pluggable modules for metro solution have gone through C-form factor pluggable (CFP) to CFP2 and future quad small form-factor pluggable double density (QSFP-DD) and Octal Small Formfactor Pluggable (OSFP) via multi-source agreement (MSA) standardization for smaller footprint, lower cost, and lower power dissipation. However, the solution is still overengineered, too expensive, too big, or too power hungry for the access. It is not an efficient or practical implementation for access applications [5].

An access network is a totally different environment as compared with long haul and metro. Typically, it includes point-to-point (P2P) and point-to-multipoint (P2MP) architectures for edge aggregation and direct connectivity to end users. To reduce the power consumption and thereby meet the size and cost requirements for access applications, the development of both low-complexity ASIC and optics is essential. In particular, the co-design of ASIC and optics to trade performance against complexity, cost, and power consumption is imperative [6].

Apart from the natural technology advancement of coherent optics and the reuse of existing fiber infrastructure, there are three major factors that influence cost and power reduction and eventually lead to a successful deployment in access networks. As shown in Figure 4.1, the first factor is the optimization in access networks – the digital, optical, and electrical complexity must be minimized in order to reduce both manufacturing costs and operational power consumption. Another important factor to consider is the standardization and interoperability in order to provide an organic ecosystem and simplify the deployment of multi-vendor networks. The final factor is the volume as the economies of scale in access are much larger than long haul or metro and can also align with data center–interconnection applications.

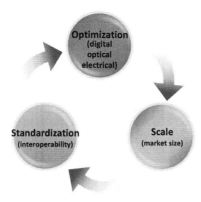

FIGURE 4.1 Three major factors for impacting coherent optics in access networks.

4.2.1 DSP OPTIMIZATION

The current development of ASICs for DSP follows two migration directions: one is to have a programmable and comprehensive coherent DSP that is capable of processing data rates from 100 to 800G per single wavelength, with the support of higher modulation formats such as 32 or 64-quadrature amplitude modulation (QAM) and FEC with high net coding gain. The second direction is to optimize functionalities, reduce power consumption, and thereby meet the size and cost requirements for access applications.

In access applications, a shorter transmission reach induces less distance-dependent signal degradation, less required link equalization (i.e. fewer digital filter taps), and less DSP processing for impairment compensation, such as CD compensation, PMD compensation, and polarization tracking. A reduction in optical signal-to-noise-ratio performance is acceptable for shorter reach access applications, which allows for a lower sampling rate and resolution of ADCs/DACs and fewer bits to be carried through the DSP. Because of shorter distance and less demand on the link budget, soft-decision FEC (SD FEC) encoder and decoder, the major blocks in terms of power dissipation of ASIC, can also be significantly simplified by either using hard-decision FEC (HD FEC), less overhead SD FEC, and/or decreasing the number of iterations. Figure 4.2 shows our analysis on the power consumption of typical ASIC from long haul, metro, to access applications. In addition to the percentage change of each constituent element, the total energy consumption is also significantly reduced.

4.2.2 OPTICAL AND ELECTRICAL COMPONENTS SIMPLIFICATION

Both optical sources, usually lasers as transmitter and local oscillator, are crucial building blocks to optimize the system cost and performance. Low-cost, small-footprint lasers with relatively large linewidth are preferred over costly narrow-linewidth external cavity lasers. In this context, lower cost lasers may provide an acceptable degradation in system performance. As shown in Figure 4.3, the analysis

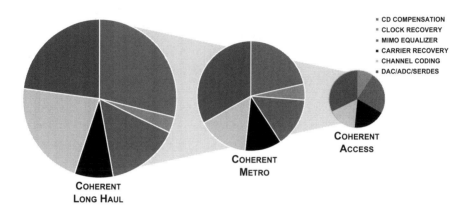

FIGURE 4.2   Analysis on power consumption of typical ASICs.

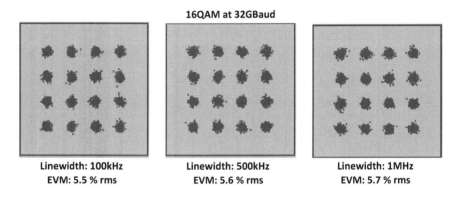

FIGURE 4.3   Impact of linewidth (phase noise) on 16-QAM signal performance.

shows little impact on 16-QAM signals when the linewidth is increased from 100 kHz to 1 MHz. This will allow the use of inexpensive laser sources for coherent access systems. Nevertheless, higher order modulation formats such as 32- or 64-QAM are more sensitive to phase noise [7].

Moreover, the cost of transceivers close to the user side should be in particular reduced in an access network. This is due to the high volume, large scale, and vast distribution of users. As a result, low-cost laser sources are preferred in these transceivers. One way to reduce the cost of lasers close to the user side is to use inexpensive Fabry–Perot lasers and lock them as slaves to the master lasers from the hub side by using injection locking. By doing this, it does not only reduce the laser cost but also produce identical wavelength for colorless operation.

A miniaturized optical I/Q modulator is another key component in a coherent optical transceiver. Being different from an intensity modulation, an I/Q modulator can include phase information onto the modulated lightwave. However, it also needs more complicated bias control and monitor. The function of an I/Q modulator can theoretically be achieved by cascading an intensity modulator and a phase modulator. Cascading two simple modulators can further reduce the complexity and the cost of a coherent optical transmitter. Of course, synchronization between the intensity modulator and the phase modulator is a must; pre-equalization of the complex signal is also necessary in order to make the cascade modulation work in the same way as an I/Q modulator.

After modulation, a booster optical amplifier, either semiconductor optical amplifier (SOA) or EDFA, is often required to compensate the coupling and modulation loss of the optical modulation process in order to provide the power budget for long-haul or metro applications. However, the cost of such optical amplifiers can be reduced because an optical link in an access environment demands a less stringent power budget.

Using low-bandwidth electrical and opto-electronic components, including driver, modulator, photodetector, and transimpedance amplifier (TIA), is another approach to reduce the overall cost in access networks. With these components operating beyond their nominal bandwidth, inter-symbol interference (ISI) or joint ISI and inter-channel interference (ICI) symbol correlation impairments will be induced. However, these impairments can be equalized in the DSP section. This is also an effective tool to balance between the cost of components and the DSP complexity.

Similarly, the sampling rate of ADCs can also be reduced to lower the cost of optical receivers. The sampling rate of an ADC is a key factor that determines the complexity, power consumption, and cost of the circuit. Ideally, the sampling rate should be higher than the Nyquist rate of the signal by some over sampling rate. However, with advanced DSP technology, sub-Nyquist sampling, i.e. sampling with a rate lower than the Nyquist rate, can also recover the original signal with acceptable degradation. In addition, using multiple parallel ADCs can significantly reduce the sampling rate of each ADC. For example, by interleaving four parallel ADCs, the rate of each ADC can be reduced by four times. This is a useful technique for high-bandwidth coherent optical receiving [8].

By leveraging the great potential of photonic integrated circuit (PIC) technology, today's coherent systems that heavily rely on discrete components have a great potential for improvement in terms of cost, power consumption, and form factor reduction. The last decade has witnessed a significant growth in the field of photonic integration. For example, today's Indium-Phosphide (InP)–based PIC allows a wafer level integration of a variety of active and passive optical components. Critical parts such as narrow linewidth lasers, SOAs, I/Q modulators, and photodetectors can be monolithically fabricated on a single wafer, allowing significant reduction in optical module cost, size, and power consumption compared to its discrete counterparts. Another example is Silicon Photonics (SiPh), a widely recognized key technology for next-generation optical communication systems. Even though monolithic integration of light sources is more challenging than InP, other key components, including modulators and detectors, have been demonstrated with good performance. However, benefits from its high index contrast and CMOS-like fabrication processes result in SiPh technology to achieve high level of integration and low manufacturing cost for high volume production. To date, PIC-based optical modules on InP and SiPh platforms have been offered extensively in commercial products. Efforts for further improvement in performance and reduction in cost and form factor continue. More details about PIC are discussed in Chapter 8.

4.2.3 STANDARDIZATION AND INTEROPERABILITY

Another important factor to consider is standardization and interoperability. Standardization in coherent optics community is mainly driven by short-reach metro/aggregation applications, where optical performance is not a differentiator. Today, around seven DSP solutions from different companies are offered. The standardization will eventually lead to improved interoperability and predictable performance and allow operators to utilize the optical fiber infrastructure more efficiently to meet future bandwidth demand.

During the standardization process, there are numerous variables to be considered to guarantee the performance and interoperability of coherent transceivers:

1. Modulation format. Being enabled by the use of a DAC at the transmitter, multiple types of modulation formats may be supported. Specified formats currently include QPSK, 8-QAM, and 16-QAM.
2. Framing format. The client signal type shall be 100 GE over a CAUI interface per IEEE 802.3ba to support most of the routers at the cable Hub.
3. DSP algorithms. The major part will be the channel encoding and decoding processes of either SD FEC or HD FEC, the symbol rates, and pulse shaping. In the case of data-aided process, the training sequences also need to be agreed upon between the Tx and Rx, such as pilot tones.
4. Symbol mapping. Gray coding minimizes the bit error rate by ensuring that the nearest neighboring symbols have a Hamming distance of 1. The constellation is Gray coded for 16-QAM, but different options exist. 8-QAM only has quasi-Gray coding because in theory, it cannot be perfectly Gray coded in theory.
5. Optical properties. To achieve interoperability, optical parameters must be defined within a strict range of working conditions. The minimum output power and its stability at the transmitter (i.e. launch power) and the sensitivity at the receiver (channel sensitivity in single channel and DWDM cases) require dedicated standardization.
6. Equalization parameters. Capability to address major fiber impairments must be specified, such as minimum CD compensation and PMD compensation requirements.

The standardization of coherent optics in access networks has mainly been led by CableLabs®. In 2017, CableLabs® announced the launch of P2P coherent optics specification-development activities. On June 29, 2018, CableLabs® for the first time publicly unveiled two new specifications: "P2P Coherent Optics Architecture Specification" and "P2P Coherent Optics Physical Layer v1.0 Specification" [9]. On March 12, 2019, CableLabs announced another addition to its family of P2P coherent optics specifications: "P2P Coherent Optics Physical Layer v2.0 Specification" which defines interoperable P2P coherent optical links running at 200 Gbps over single wavelength [10]. These specifications result from collaboration among CableLabs®, operators, manufactures, and their common interests in expanding the capacity of access networks. The industry as a whole benefits from a successful standardization of the coherent transceiver, including both optical performance and DSP functions and capabilities.

4.2.4 ECONOMIES OF SCALE IN ACCESS NETWORK

In Chapter 1, the current state of fiber deployment in telephony, cellular, and cable networks is discussed. Even though there has already been a significant deployment of fiber, it is worth considering how much additional fiber is likely to be deployed in the future as service speeds and aggregate capacities continue to grow. The amount of fiber deployed grows significantly as new fiber cable branches out to deeper points in the network and much closer to the subscriber. A closer look at the United States cable network is illuminating to assess the potential growth of coherent optics links and economies of scale for components used in the access. Today in the United States, there are about 300,000 cable fiber nodes connected with fiber to the hub and covering on average more than 400 households. Several operators are segmenting this original fiber node–serving area into multiple smaller fiber node–serving areas. One segmentation strategy being followed is known as N+0, the name being indicative in having 0 radio frequency (RF) amplifiers following the fiber node. This segmentation results in 10–18 child nodes from the original node. A total capacity of 10 Gbps per fiber node is possible leveraging today's node RF bandwidth of 1 GHz, resulting in a potential aggregate capacity of 100–180 Gbps in this residential use case. If 50 percent of the United States fiber nodes would evolve to this architecture, 300,000 coherent transceivers would be required. These are initial aggregation scenario candidates for a coherent optics termination point. The number of cable fiber nodes worldwide is estimated to exceed 1.3 million, which is bound to evolve, following some degree of node segmentation.

Demand for even higher capacity increases with time, and the use of coherent optics aggregation links is supplemented with direct end-to-end connectivity to the higher consumption end-points. Increasing RF bandwidth or segmenting nodes further are future evolution alternatives. Higher demand for capacity leads to a dedicated coherent link to the child nodes. Enterprise customers and cellular base stations, expected in a 5G deployment scenario, may also have greater demand for capacity. In the United States, there are more than 600,000 businesses with more than 20 employees [11]. A sizeable portion of these may require Gbps services in the near future. In addition, the numerous small cells may result from cellular evolution to 5G. All these access use case scenarios favorably contribute to the economies of scale of coherent optics components through either aggregation or direct end-to-end connectivity. The access network is the largest component of the network in terms of physical/geographic scale and investment. Along with increasing shipment volume in long haul and metro, the access will help drive coherent optics component and equipment pricing down.

REFERENCES

1. H. Sun, K.-T. Wu, and K. Roberts, "Real-time measurement of a 40 Gb/s coherent system," *Optics Express*, 16(2), 873–879, 2008.
2. Z. Jia, J. Yu, H.-C. Chien, Z. Dong, and D. Di Huo, "Field transmission of 100 G and beyond: multiple baud rates and mixed line rates using Nyquist-WDM technology," *Journal of Lightwave Technology*, 30(24), 3793–3804, 2012.
3. H. Rohde, S. Smolorz, S. Wey, and E. Gottwald, "Coherent optical access networks," Optical Fiber Communications Conference, paper OTuB1, 2011.
4. A. Teixeira, A. Shahpari, R. Ferreira, F. P. Guiomar, and J. D. Reis, "Coherent access," Optical Fiber Communications Conference, paper M3C.5, 2016.
5. E. Wong, "Next-generation broadband access networks and technologies," *Journal of Lightwave Technology*, 30(4), pp. 597–608, 2012.
6. L. N. Binh, *Digital Processing: Optical Transmission and Coherent Receiving Techniques*, CRC Press, Boca Raton, FL, 2013.
7. J. Pfeifle, and M. Koenigsmann, "Enabling 400G/1T coherent communications," Keysight Webcast, 2016.
8. W. Jiang, K. G. Kuzmin and W. I. Way, "Effect of low over-sampling rate on a 64Gbaud/DP-16QAM 100-km optical link," *IEEE Photonics Technology Letters*, 30(19), 1671–1674, 2018.
9. "P2P Coherent Optics Physical Layer 1.0 Specification – P2PCO-SP-PHYv1.0-I02-190311," 11 March 2019, Cable Television Laboratories, Inc.
10. "P2P Coherent Optics Physical Layer 2.0 Specification – P2PCO-SP-PHYv2.0-I01-190311," 11 March 2019, Cable Television Laboratories, Inc.
11. US Small Business Adnministration, "2018 Small business profile," www.sba.gov/sites/default/files/advocacy/2018-Small-Business-Profiles-US.pdf

5 Coherent Optics Use Cases in Access Networks

Chris Stengrim and Junwen Zhang

CONTENTS

5.1 INTRODUCTION

When the technical performance of coherent optics vs. alternative technologies proves superior, the economics often determine whether to deploy coherent optics. While many use cases in the access network exist, scenarios requiring a high bandwidth will highlight the technical and economic superiority of coherent optics, i.e., well economically coherent optics scales with higher capacity applications.

5.2 DEPLOYMENT SCENARIOS

Coherent optics in the access network can have two deployment scenarios. The first is the aggregation of multiple optical links between a hub and edge devices (or child nodes) on the network. These edge devices, in turn, connect to end points, such as residences, multi-dwelling units, enterprises, data centers, or wireless access points. The second is the direct end-to-end (E2E) connectivity from the hub to the end points.

5.2.1 AGGREGATION USE CASE

Cable operators, telcos, and even mobile network operators have wireline networks extending to multiple network edge devices. These networks may have optical fiber, copper, coax, or a combination of these physical mediums. Example of network edge devices includes hybrid fiber-coaxial (HFC) nodes in cable networks, digital subscriber line access multiplexer (DSLAM) 's in telco networks, Local Convergence Points in fiber-to-the-home (FTTH) networks, and wireless base stations or small cells in mobile networks. Figure 5.1 illustrates many different edge devices with optical links consolidated at an aggregation node and transmitted on a single optical link to a hub.

5.2.2 DIRECT END-TO-END CONNECTIVITY USE CASE

The second use case is for E2E services. For this use case, there is a dense wavelength division multiplexing (DWDM) multiplexer (Mux) at the hub that combines multiple Point-to-Point (P2P) coherent optic links onto a fiber. At the aggregation node, another DWDM Mux splits the P2P coherent wavelengths and puts each on its own fiber strand. The de-multiplexed P2P coherent optic Links will connect directly to the end point. The P2P coherent optic links from the hub could be a mix of 100- and 200-Gbps link rates. For this use case, the P2P coherent optic transceiver will exist in a device at the end point; therefore, the P2P coherent optic link distance is from the hub all the way to the end point. There could still be a mix of optical link types coexisting with the E2E coherent links including analog or non-coherent links within the same fiber.

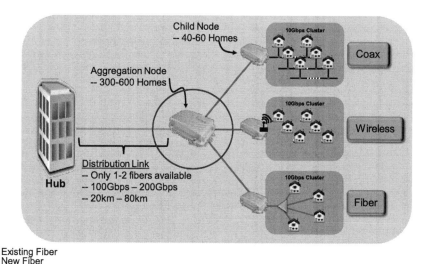

FIGURE 5.1   10G IM-DD DWDM aggregation of multiple child nodes with RPDs.

5.3 EXAMPLE: CABLE ACCESS NETWORKS

5.3.1 Cable HFC Network

The Cable Network's evolution to a Distributed Access Architecture (DAA) naturally fits into an aggregation use case. Cable operators have begun evolving their access networks from centralized architectures to distributed architectures. In a centralized access architecture, the Cable Modem Termination System (CMTS) or Converged Cable Access Platform (CCAP) resides in a hub providing broadband connectivity to thousands of cable modems across dozens or even hundreds of nodes. Distributed architectures are arranged in a variety of ways: in Remote physical layer (PHY), media access control (MAC) functionality remains in the hub, while PHY functionality is pushed deeper into the access network; in Remote MAC–PHY, both MAC and PHY functionalities are pushed deeper into the access network; and in other emerging topologies, flexible placement means MAC functions can reside either in a hub or deeper in the network.

These DAAs require the placement of fiber beyond the current HFC node to reach child nodes deeper in the network that house Remote PHY Device (RPD) or Remote MAC–PHY device. In the case of Remote PHY, each RPD must connect to the CCAP core at the hub through a digital optical link. The existing fiber links between the hub and the distributed devices need to be upgraded from analog-to-digital optics in order to support distributed architectures.

Typical HFC network designs use 6–8 fibers to connect the hub to the fiber node. Two of these fibers are used for primary downstream and upstream connections, and in some cases, two additional fibers are used for redundancy. The rest of the fibers are included for future use. Unfortunately, many of these "future-use" fibers have since been repurposed for business services, cell tower backhaul, and node splits. In some cases, only the two primary fibers feeding the fiber node remain available for access transport. Overbuilding existing fiber links from hub to node and deeper to reach new child nodes can be cost-prohibitive, so cable operators have used digital optical technologies to improve the capacity of existing fiber. Coarse wavelength division multiplexing (CWDM) and DWDM technologies enable multiple wavelengths (λ) on a single fiber to expand the capacity linearly.

Intensity-modulation and direct-detection (IM-DD) is a common digital optical technology deployed today by cable operators. Multiple 10G IM-DD optical links can be multiplexed by using DWDM. For example, as shown in Figure 5.2, a DWDM multiplexer expands the fiber capacity by enabling 16 wavelengths of 10 Gbps capacity each to be deployed on the existing fiber links from the hub to the aggregation node (this could be the same location as the current HFC node). There are DWDM Muxs supporting even higher quantity wavelengths, such as 40 or 44 wavelengths, but they are not necessary for the use case that follows.

The 10G IM-DD DWDM solution works when scaling a selected number of child nodes, but there are some inefficiencies. When multiplexing a large quantity of wavelengths or transmitting over a longer distance, optical amplification and chromatic dispersion (CD) compensation will be required on the optical link. With IM-DD, each child node (with an RPD) requires a dedicated wavelength to provide 10-Gbps

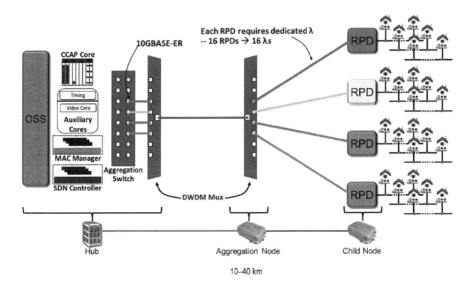

FIGURE 5.2 10G IM-DD DWDM aggregation of multiple child nodes with RPDs.

capacity. Wavelengths become scarce when aggregating multiple child nodes and meeting demands for other fiber-based services. Colored DWDM optics are required for each link, resulting in higher costs compared to gray optics, as described later.

With a 10G IM-DD DWDM solution, the optical transceiver must support the entire distance from the hub to the child node. As shown in Table 5.1, there are three 10GBASE standards, each with a different distance limitation. Any span over 10 km requires either the 10GBASE-ER or 10GBASE-ZR transceiver, which may incur higher network costs compared to the 10GBASE-LR transceiver.

Coherent optics addresses the inefficiencies of IM-DD with DWDM and provides greater technical performance. It enables a capacity of 100 Gbps or higher on a single wavelength at a distance up to 40 km without amplification. In the use case shown in Figure 5.3, coherent optics uses a single wavelength to aggregate the traffic from the hub to the node. With the aggregation node deployed 10–40 km from the hub, the shorter distance from the child nodes to the aggregation node allows the use of a 10GBASE-LR transceiver to connect them.

TABLE 5.1

10GBASE Standards

Standard	Model	Data Rate	Distance (km)	Fiber Mode	Wavelength (nm)
10GBASE-LR	SFP-10G-LR	10G	10	Single	1,310
10GBASE-ER	SFP-10G-ER	10G	40	Single	1,550
10GBASE-ZR	SFP-10G-ZR	10G	70–80	Single	1,550

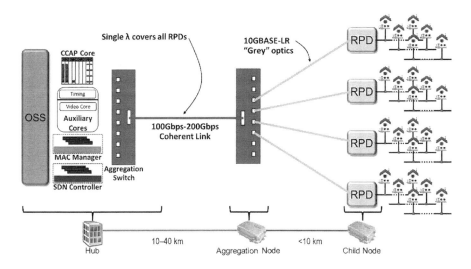

FIGURE 5.3   Coherent optics aggregation of multiple child nodes with RPDs.

5.3.2   Economic Model

An economic analysis using an actual North American cable node considers a similar use case of optical aggregation in a DAA. This use case assumes that the distance from the hub to the aggregation node is less than 40 km. From Figures 5.2 and 5.3, the digital optical link from the hub to the aggregation node could use either $N \times 10G$ IM-DD DWDM optical links or a 100G P2P coherent optics link. The aggregation node could use either a switch, a router, or a muxponder to transmit to and receive from the child nodes. For this use case, a 100G muxponder can allocate up to 10×10 Gbps lanes for communication with the child nodes. There are pros and cons for using a switch or a router instead of a muxponder, but those comparisons are beyond the scope of this chapter.

The economic analysis uses a proprietary methodology to consider the component costs and wavelength costs for the aggregation of 1–10 child nodes. The 10G IM-DD DWDM solution may or may not require the use of optical amplifiers, depending on distance and the number of wavelengths in the Mux. Rather than showing all the variations in the optical link budget, the analysis considers the 10G IM-DD DWDM solution with and without amplifications.

This analysis utilizes publicly available unit costs provided by third-party research analysts and optical component costs listed on distributor websites. Unit costs available to cable operators can vary significantly, so the results of the analysis might not apply to all operators. As shown in Figure 5.4, for this use case, 10G IM-DD DWDM proves more cost-effective when aggregating up to 6 or 7 child nodes, depending on the need for optical amplification. Coherent optics proves more cost-effective in this use case when aggregating 7–10 child nodes.

There are several variations of the aggregation use case to consider, and coherent optics provides economic benefits in many of these cases, especially at scale. The use

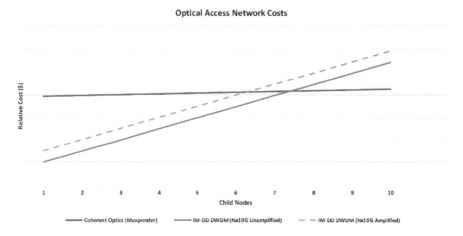

FIGURE 5.4 Cost comparison of coherent optics (with Muxponder) to IM-DD with and without amplification.

of a switch or router can provide aggregation of more than 100 Gbps of capacity by using oversubscription. Additionally, the development of 200G P2P coherent optics technologies for the cable access network will provide even greater scale to connect a growing number of optical links. As cable operators connect not only their core residential broadband networks but also their enterprises and wireless networks, 200G P2P coherent optics technologies will become a necessity. Over time, the economics of coherent optics will continue to improve as optical vendors optimize their components for cable access networks.

5.4 COHERENT PON (P2MP)

The advance of high-speed wired access network has been propelled by new business and technology drivers, such as cloud services, 5G wireless transport, and high bandwidth 4/8K video applications. The bandwidth requirements in the access network are expected to grow to Gb/s for residential offerings and multi-Gb/s for business markets in near-future optical access networks. As a point-to-multipoint (P2MP) system, passive optical network (PON) technologies have been one of the dominant architectures to meet such high capacity demand for the end users [1–3]. This is further evidenced by the progress in standards bodies developing next-generation high-speed time-division multiplexing PON (TDM-PON) standards. The IEEE 802.3ca 100G Ethernet PON (EPON) Task Force is moving toward the standardization of 25/50G EPON based on wavelength multiplexing of 25 Gbps per single channel [1]. New projects to standardize higher speed PONs, e.g., 50G, has also been started in the ITU-T Q2/SG15 group [2].

High-speed PON based on single wavelength with TDM mechanism proves to be the most attractive solution, which not only reduces the number of required optical components and the associated cost but also saves the wavelength resources.

However, limited sensitivity has become a critical challenge to support high-speed PON with high power budget using direct detection.

Generally, coherent detection is an effective method to increase the receiver sensitivity [4–6]. There are many advantages of using coherent detection in a PON: it largely improves the receiver sensitivity gained by coherent beating of signal with clean local oscillator (LO) signal, and therefore, it can support longer distance transmission and/or larger number of end users; it also enables high access speed due to the multi-dimensional and more advanced modulation formats with higher spectral efficiency. Coherent detection also enables digital signal processing (DSP), which makes optical and electrical impairments, e.g., CD, to be fully handled in the digital domain. As discussed in [6], bi-directional 25G/50G TDM-PON with extended power budget is enabled by coherent detection to provide an additional gain in upstream. In [7], the authors demonstrated a 4×25-Gb/s coherent ultra DWDM (UDWDM) PON over 50-km standard single-mode optical fiber, in which the downlink power budget can be more than 44 dB to support more than 1,000 subscribers. Therefore, coherent technologies have been recently regarded as one of the most future-proof candidates for next-generation optical access networks.

As described in Section 2, Coherent PON is also used in the two deployment scenarios: the aggregation use case and direct E2E connectivity. Similar to other PON architectures, Coherent PON would comprise an Optical Line Terminal in the hub and Optical Network Units (ONU). The placement of the ONU would depend on the deployment scenario. In the aggregation use case, the CPON ONU would be connected to the edge devices in lieu of 10G optics. In the direct E2E connectivity use case, the ONU would naturally reside at the end point – similar to current PON deployments. The development of Coherent PON is still nascent, with a commercially available solution yet a few years away from reality.

5.5   CHALLENGES IN COHERENT PON FOR ACCESS NETWORK

Despite having a number of benefits from coherent technology, there exist several challenges of introducing coherent PON into access network. Cost is the first challenge. Access network, which is directly connected to end users, is very cost-sensitive due to its large-scale market size. As the coherent technology was first used in a P2P long-haul optical system which utilizes best-in-class discrete photonic and electronic components, it cannot be directly implemented in Coherent PON due to the high cost. Laser, modulators, and detectors all need additional optimization to lower the cost. Recently, many efforts have been made by using different kinds of simplification process to resolve cost issue [6]. In [8], the authors demonstrated single-wavelength 100-Gb/s four-level pulse amplitude modulation (PAM-4) TDM-PON transmission in C-band using simplified and phase-insensitive heterodyne coherent detection, with a power budget over 32 dB. Chapter 6 will discuss the recent progress in simplified coherent optical solutions.

Upstream burst detection is the second challenge or the second series of challenges for coherent PON to be implemented [9]. Unlike downstream, upstream in coherent PON is the case of "from many to one". The upstream signals are from different users, they may have different wavelength or frequency offset, totally different clock, phase, and polarization statues. As analyzed in [9], a burst laser transmitter can cause large frequency or wavelength drift. Different from direct detection, the signal wavelength should be locked to or very close to the LO laser in the coherent detection; otherwise, the signal would be out of the detection bandwidth. One solution, which should not be the only one, has been reported including self-coherent detection and semiconductor-optical-amplifier-based reflection and modulations [10].

The upstream signals may have different signal strength, as they may come from different end users in different locations. Therefore, another big challenge of the burst detection is the burst signal amplification. In current non-coherent PON system, e.g., 10G EPON or XGs-PON, the burst amplification is generally realized in the electrical domain using burst-mode limiting transimpedance amplifier (TIA). However, it becomes challenging for high-speed burst signals over 10 Gb/s. It is even more challenging since coherent upstream burst receiver not only requires multiple burst TIAs for different data paths but also demands the TIA to be linear and identical in each path, i.e., four burst linear TIA for dual-polarization coherent receiver. Recently, a 25-GHz burst linear TIA is reported in [11], where the authors presented a linear burst-mode TIA capable of receiving 50 Gb/s PAM-4. These results show the possibility for future burst linear TIA in any coherent detection systems.

Finally, burst DSP is yet another challenge to be solved in coherent PON. For downstream with continuous mode, the system has all the time in the world to react, prepare, and decide how to detect the signal. Most of the reported DSP algorithms are designed for P2P links in continuous mode [12]. As mentioned previously, the receivers should adjust the received signals from different frequency offset, totally different clock, phase and polarization statues with its DSP statues every time a signal comes from a different end-device. The adjustment should be in a very fast way in order to efficiently demodulator the signal. In non-coherent PON, such as 10G EPON, the maximum allowed clock and data recovery time is 400 ns [13]. It is expected that the burst signal recovery time should also be less than 400 ns, which makes many reported algorithms on blind carrier recovery, i.e., bland frequency offset estimation and phase recovery, not suitable [14–15]. Some blind channel estimation methods also need a special design to achieve fast convergence [5]. However, many reports [16–18] believe that burst signal processing can be achieved with special digital signal-processing designs.

5.6 SUMMARY

The high bandwidth enabled by coherent optics in the access network requires the exploration of use cases to take advantage of the technical superiority compared to other optical technologies. High bandwidth users, such as enterprises, network operators, and data centers, may require such high capacity on their own. More likely for access-network operators is the aggregation of multiple end points into a single large pipe benefiting from the scale provided by coherent optics. The aggregation links can be provided by P2P Coherent Optics today and may evolve into P2MP Coherent Optics in the near future. The motivations and challenges of Coherent PON in access are also discussed in this chapter with the most recent research progresses.

REFERENCES

1. V. Houtsma et al., "Recent progress on standardization of next generation 25, 50 and 100G EPON," *Journal of Lightwave Technology*, 35(5), pp. 1228–1234, 2017.

2. ITU-T Q2/SG15, "Proposal for the study of 50G TDM-PON," ITU-T SG 15, Contribution 641, 2017.

3. T. Minghui, et al., "50-Gb/s/λ TDM-PON based on 10G DML and 10G APD supporting PR10 link loss budget after 20-km downstream transmission in the O-band," 2017 Optical Fiber Communications Conference and Exhibition (OFC), Los Angeles, CA, p. 2, paper Tu3G.2, 2017.

4. A. Shahpari, et al., "Terabit + (192×10 Gb/s) Nyquist shaped UDWDM coherent PON with upstream and downstream over a 12.8nm band," Optical Fiber Communication Conference/National Fiber Optic Engineers Conference 2013, paper PDP5B.3, 2013.

5. K. Matsuda, et al., "Hardware-efficient adaptive equalization and carrier phase recovery for 100-Gb/s/λ-based coherent WDM-PON systems," *Journal of Lightwave Technology*, 36(8), pp. 1492–1497, 2018.

6. D. van Veen and V. Houtsma, "Bi-directional 25G/50G TDM-PON with extended power budget using 25G APD and coherent amplification," Optical Fiber Communication Conference Postdeadline Papers, paper Th5A.4, 2017.

7. M. Luo, et al., "Demonstration of bidirectional real-time 100 Gb/s (4×25 Gb/s) coherent UDWDM-PON with power budget of 44 dB," Optical Fiber Communication Conference (OFC) 2019, OSA Technical Digest, Optical Society of America, paper Th3F.2, 2019.

8. J. Zhang, et al., "Single-wavelength 100-Gb/s PAM-4 TDM-PON achieving over 32-dB power budget using simplified and phase insensitive coherent detection," European Conference on Optical Communication (ECOC), Rome, pp. 1–3, 2018.

9. V. Houtsma and D. van Veen, "Optical strategies for economical next generation 50 and 100G PON," Optical Fiber Communication Conference (OFC), OSA Technical Digest, Optical Society of America, paper M2B.1, 2019.

10. D. Kim, et al., "80-km reach 28-Gb/s/λ RSOA-based coherent WDM PON using Dither-frequency-tuning SBS suppression technique," Optical Fiber Communication Conference (OFC), OSA Technical Digest, Optical Society of America, paper Th3F.6, 2019.

11. G. Coudyzer, et al., "A 50 Gbit/s PAM-4 linear burst-mode transimpedance amplifier," *IEEE Photonics Technology Letters*, 31(12), pp. 951–954, 2019.

12. S. J. Savory, "Digital coherent optical receivers: Algorithms and subsystems," *IEEE Journal of Selected Topics in Quantum Electronics*, 16(5), pp. 1164–1179, 2010.

13. IEEE 802.3av-2009, Institute of Electrical and Electronics Engineers, 2009, https://standards.ieee.org/standard/802_3av-2009.html

14. E. Ip and J. M. Kahn, "Feedforward carrier recovery for coherent optical communications," *Journal of Lightwave Technology*, 25, pp. 2675–2692, 2007.

15. Y. Gao, et al., "Low-complexity and phase noise tolerant carrier phase estimation for dual-polarization 16-QAM systems," *Optics Express*, 19, pp. 21717–21729, 2011.

16. B. C. Thomsen, et al., "Burst mode receiver for 112 Gb/s DP-QPSK with parallel DSP," *Optics Express*, 19, pp. B770–B776, 2011.

17. M. Li et al., "Optical burst-mode coherent receiver with a fast tunable LO for receiving multi-wavelength burst signals," Optical Fiber Communications/National Fiber Optics Engineers Conference, Los Angeles, CA, pp. 1–3, 2012.

18. R. Koma, et al., "Wide dynamic range burst-mode digital coherent detection using fast ALC-EDFA and pre-calculation of FIR filter coefficients," Optical Fiber Communications Conference and Exhibition (OFC), Anaheim, CA, pp. 1–3, 2016.

6 Simplified Coherent Optics for Passive Optical Networks

Domanic Lavery and Mustafa Sezer Erkilinc

CONTENTS

6.1 INTRODUCTION

Passive optical networks (PONs) are based on transmission using intensity-modulated lasers and direct-detection receivers as has been the case since their development over two decades ago. Access networks have recently introduced line rates up to 25 Gb/s but are now facing the limitations of bandwidth scaling when using intensity-modulation and direct-detection (IM-DD). To temporarily bypass this issue, standards have now resorted to wavelength-division multiplexing (WDM). Beyond this, the IEEE EPON Working Group 802.3ca [1] is exploring options for advanced, soft-decision, forward error correction (FEC) codes [2–3], and discussions are ongoing regarding high-speed avalanche photodiodes, multilevel modulation, external modulation, and even digital signal processing (DSP) for feed-forward equalization of channel and transceiver impairments. However, there is a marked reluctance to move away from IM-DD, which is the limiting element of the communication system.

The primary driver of this intransigence is cost (see Chapter 4, Section 2). Although the well-established alternative, coherent detection, has now seen a decade of use in long-haul transmission systems, the cost and complexity will need to be substantially reduced to enable adoption in optical access networks. This chapter discusses options for simplified coherent receiver transceiver designs and provides insights into the implications for DSP complexity.

6.2   OPTICAL AND ELECTRICAL COMPONENTS SIMPLIFICATION

Two electromagnetic waves are said to be coherent if the relative phase between them remains constant. The receiver in which a signal- and a local-phase reference (local oscillator (LO) laser) are combined for detection can be said to be a coherent receiver – a selection of their designs is shown in Figure 6.1a. In optical fiber communication systems, typically polarization- and phase-diverse *intradyne* reception, that is, using two separate (transmitter and LO) lasers operating at a comparable wavelength, is commonly used [4], as shown in Figure 6.1a. Alternatively, the optical field can be detected as a purely real-valued signal using *heterodyne* detection, shown in Figure 6.1b, which halves the number of required optical hybrids, analog-to-digital converters (ADCs), and balanced photodiodes required, at the expense of doubling the optoelectronic bandwidth requirements and an inherent 3 dB sensitivity penalty.

To achieve any further simplification, the removal of polarization diversity, as per Figure 6.1c, is required. However, to operate in practice, the signal must be pre-coded across two polarizations such that the LO state of polarization is guaranteed to match the state of polarization of the transmitted signal. Two practical methods for this are Alamouti coding [5], which is a half-rate polarization-time block code, and polarization scrambling [6], which switches the signal state of polarization between two orthogonal states twice per symbol period.

In contrast to IM-DD receivers, which are limited by thermal noise and typically operate at 20–25 dB [7] from the fundamental sensitivity limits, coherent detection systems are limited by shot noise and can approach single-photon/bit sensitivity. This is due to the relatively high LO power compared to the received signal power [8]. This difference is illustrated in Figure 6.2, where the sensitivity of receivers (shown in Figure 6.1a–c) is compared to a comparable IM-DD system, assuming 50 Gb/s per channel.

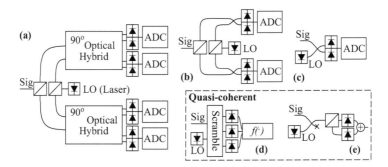

FIGURE 6.1   (a) A polarization- and phase-diverse coherent intradyne receiver, (b) a polarization diverse coherent receiver using heterodyne detection, (c) a single-polarization heterodyne coherent receiver, (d) a general model of quasi-coherent receivers, and (e) a common example of quasi-coherent detection. In (a)–(c), balanced receivers are shown, but single-ended detection is also possible. In all the cases, the incident signal is marked as "Sig".

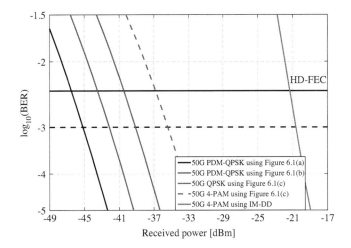

FIGURE 6.2   The sensitivity limit of coherent receivers (shown in Figure 6.1) using 50-Gb/s QPSK and 4-PAM signals, and an IM-DD receiver with an ideal photodiode.

In this analysis, the typical receiver electrical noise sources, for example, ADC quantization noise and noise from the trans-impedance amplifiers, are ignored, and the quantum efficiency is assumed to be 1, corresponding to an ideal photodiode (PD). The sensitivity limit of a 50 Gb/s quadrature phase-shift-keying (QPSK) signal detected using a polarization- and phase-diverse coherent receiver (limited only by shot noise) is found to be −46.4 dBm at a hard-decision FEC bit-error ratio of 4×10^{-3}. The sensitivity limits degrade to −43.4 and −40.4 dBm after performing heterodyne detection using the receiver design, as shown in Figure 6.1b, and single-polarization signal coherent reception, as shown in Figure 6.1c, respectively. If one chooses to reduce the coherent transmitter complexity in addition to receiver complexity by using 4-*ary* pulse amplitude modulation (4-PAM), instead of QPSK, a sensitivity penalty of 3.6 dB is observed, resulting in −36.8 dBm sensitivity. However, a sensitivity of only −21.3 dBm can be achieved using an ideal IM/DD transceiver employing a 50-Gb/s 4-PAM signal. Following these simplifications in a coherent transceiver architecture, it is found out that there is still a 15-dB sensitivity margin compared to an IM-DD system, which can be utilized to further relax the optical front-end requirements or the required DSP complexity.

An additional class of the simplified receiver, known as quasi-coherent receivers, combines optical network unit (ONU)–side polarization scrambling (using a polarization beam splitter (PBS)) with envelope detection, as shown in the general configuration in Figure 6.1d. These receiver configurations can be operated without DSP [9–10] while preserving most of the sensitivity gain of coherent detection. However, they are not compatible with chromatic dispersion compensation (CDC) or advanced modulation formats, due to the loss of phase diversity. For this reason, it can be challenging to increase the bit rate per wavelength. A promising quasi-coherent design, which effectively extends the reach of 10 Gb/s PONs without optical amplification, is depicted in Figure 6.1e [11].

6.3   OPTIMIZED DIGITAL SIGNAL PROCESSING ALGORITHMS

DSP is not necessarily a requirement for low-complexity coherent transmission systems, as discussed in the previous section. In fact, due to their nonlinear detection profile, quasi-coherent receivers generally do not significantly benefit from the addition of DSP. However, for linear coherent receivers, depicted in Figure 6.1a–c, the DSP can offer several additional advantages over direct detection. In this section, we focus on two of the most computationally expensive signal-processing algorithms, namely, CDC and adaptive equalization, and show how they can be simplified for short-reach transmission systems.

6.3.1   CHROMATIC DISPERSION COMPENSATION

In contrast to direct-detection systems, DSP can be used to perfectly compensate the inter-symbol interference caused by chromatic dispersion. The number of finite impulse response (FIR) filter taps required for CDC is given in [12]

$$N = 1 + 2 \left\lfloor \frac{|D| z c f^2}{2 v^2} \right\rfloor \tag{6.1}$$

where the dispersion parameter $D = 17$ ps/nm/km in the C-band of standard single-mode fiber (SSMF), z is the transmission distance, c is the speed of light in vacuum, f is the symbol rate, and v is the carrier frequency. In long-haul communication systems, the high symbol rates and long transmission distances lead to substantial accumulated chromatic dispersion, that is, typically on the order of thousands of symbols, requiring long FIR filters to compensate. For example, a moderate symbol rate of 50 GBd over 10,000 km would require over 13,000 filter taps to perfectly compensate for chromatic dispersion. Such complex filters are essential but dominate the complexity and power consumption of the coherent receiver DSP [13]. However, in optical access networks, the transmission distances are shorter of two orders of magnitude, and the greatest symbol rates currently under consideration are half this value. Therefore, the required number of taps (N) to compensate for the dispersion accumulating in a typical access link is significantly less than a long-haul link. The increase in the number of taps with respect to symbol rate and transmission distance, assuming a signal sampled at twice the symbol rate, is plotted in Figure 6.3. Consider a C-band PON transmission operating at 25 GBd over 40 km of SSMF, a challenging symbol rate, but with a distance commensurate with next-generation PON2 [14]. This would require only 13 FIR filter taps to compensate for the accumulated chromatic dispersion, which is a value comparable to the feed-forward equalizers currently considered for future IM-DD PONs [15].

Chromatic dispersion has an all-pass response, but this introduces a potential issue when using short filters for compensation. To create an all-pass digital FIR filter with the correct phase and amplitude characteristics, a large number of taps are required; an insufficient number of taps leads to signal distortions. However, new FIR filter-design techniques have been proposed to eliminate the filtering penalty for reasonable length FIR filters, while incurring no additional computational complexity [16].

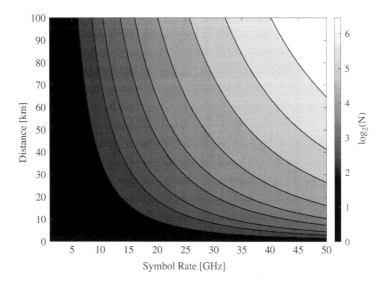

FIGURE 6.3 Chromatic dispersion compensation filter tap number, N, requirements with respect to transmission distance and symbol rate. Assumes C-band transmission over standard single-mode fiber and that the signal is sampled at twice the symbol rate.

6.3.2 ADAPTIVE EQUALIZATION

When transmitting signals over long transmission distances, time-dependent variations in the channel response, for example, the optical fiber birefringence, require adaptive channel equalization (see Chapter 3, Section 6). Even for single-span transmission, it cannot be assumed that the channel is static and, as previously noted, the received state of polarization cannot be guaranteed. Polarization-diverse coherent receivers detect the orthogonal polarization states of the optical field and use multiple-input–multiple-output (MIMO) equalization to recover the transmitted signals. This requires two parallel FIR filters per polarization (one for equalization and the other for the cross-polarization mixing). Due to the parallel FIR filters and the feedback loop required to adaptively update the tap weights, the adaptive equalizer significantly contributes to DSP power consumption.

When using a single-polarization receiver, for example, Figure 6.1c, information at one state of polarization is lost and MIMO equalization is no longer possible. As previously noted, Alamouti coding can be used to pre-code the transmitted signal such that the signal can be received, regardless of the received state of polarization. In practice, this means exchanging the polarization-based MIMO equalizer for a time-based MIMO equalizer. This has a comparable computational complexity to the MIMO DSP used in a polarization-diverse coherent receiver.

However, if the signal is instead polarization-scrambled at the transmitter, the signal can be recovered through adaptive equalization without MIMO. This is because the signal is guaranteed to align with the polarization state of the LO for at least 50% of the symbol period. Single-polarization adaptive equalization does not require the cross-polarization filters and, thus, can be implemented with one quarter of the complexity of the MIMO filter. Such low complexity equalizers can be readily implemented in real time using software-defined processing, even at Gb/s line rates [17].

There are several important consequences of using single-polarization equalizers. First, the convergence rate of coherent burst mode equalization is generally limited by polarization recovery. If the signal polarization state is already aligned with the LO, the recovery can, in principle, be significantly faster. Second, if the equalizer has sufficient filter taps to match the channel response, it can additionally be used to compensate for chromatic dispersion and apply matched filtering, if necessary, requiring no changes to the equalizer structure itself. It is possible to trade off the penalty for very short adaptive filters and the overall system performance. It has previously been shown that despite accepting a sensitivity penalty of less than 1 dB, full channel equalization can be achieved with filters as short as 5 taps [18].

6.4   SUMMARY

In this chapter, we discuss the possibility of significantly reducing the optoelectronics hardware complexity of a coherent receiver to facilitate implementation in cost-sensitive applications, such as PONs. The most compelling designs for optical access have common features. Polarization agnostic or independent detection reduces the requirements of the optical front-end by eliminating the requirement for a PBS, which would normally be required for polarization diversity. By additional use of heterodyne coherent detection, where the phase of the optical field can be recovered without an optical hybrid, the simplest coherent receiver designs can be implemented using only a single photodiode and a local laser (as a phase reference). It is important to note that this is comparable to the optical complexity of the ONUs in use in the current generation of direct-detection PONs. However, for such a simple receiver configuration to be viable in practice, polarization diversity or polarization scrambling is required at the optical line terminal transmitter to avoid signal fading at the ONU receiver.

The most simplified coherent receiver designs can operate without DSP, but this comes at the expense of sacrificing the coherent receiver's linear detection profile. Therefore, such analog (quasi-)coherent receivers require the inter-symbol interference induced by accumulated chromatic dispersion to be negligible and the received signal to be restricted to only amplitude modulation. Since the advantages of coherent receivers can only be fully exploited with DSP, this chapter therefore explores the DSP complexity requirements for coherent receivers operating in PONs.

The DSP requirements for long-haul coherent transmission systems are dominated by CDC and adaptive equalization. Here, it is shown that the DSP requirements for most PON systems are comparable to the feed-forward equalizers under the consideration for IM-DD PON. Additionally, channel equalization is also shown to have trivial computational complexity for low symbol rate, single-polarization coherent receivers.

REFERENCES

1. IEEE P802.3ca 50G-EPON Task Force, www.ieee802.org/3/ca/
2. F. Effenberger et al., "Enhanced FEC consideration for 100G EPON," 2016, www.ieee802.org/3/ca/public/meeting_archive/2016/03/effenberger_3ca_2_0316.pdf
3. V. Houtsma et al., "Impact of pre-coding and high gain FEC on 25, 50 & 100G EPON," 2016, www.ieee802.org/3/ca/public/meeting_archive/2016/09/houtsma_3ca_1_0916.pdf
4. E. Ip et al., "Coherent detection in optical fiber systems," *Optics Express*, vol. 16, no. 2, pp. 753–791, 2008.
5. M. S. Erkılınç et al., "Polarization-insensitive single-balanced photodiode coherent receiver for long-reach WDM-PONs," *Journal of Lightwave Technology*, vol. 34, no. 8, pp. 2034–2041, 2016.
6. I. N. Cano, et al., "Polarization independent single-PD coherent ONU receiver with centralized scrambling in udWDM-PONs," Proceedings of European Conference on Optical Communication, paper P.7.12, 2014.
7. G. P. Agrawal, *Fiber-Optic Communication Systems*, 4th edition, John Wiley & Sons, McGraw-Hill, 2010.
8. B. Zhang, et al., "Design of coherent receiver optical front end for unamplified applications," *Optics Express*, vol. 20, no. 3, pp. 3225–3234, 2012.
9. J. A. Altabas et al., "25Gbps quasicoherent receiver for beyond NG-PON2 access networks," Proceedings of European Conference on Optical Communication, We2.70, 2018.
10. M. Rannello et al., "Optical vs. electrical duobinary for 25 Gb/s PONs based on DSP-free coherent detection," Proceedings of Optical Fiber Communications, M1B.6, 2018.
11. J. A. Altabas et al., "Real-time 10Gbps polarization independent quasicoherent receiver for NG-PON2," Proceedings of Optical Fiber Communications, Th1A.3, 2018.
12. S. J. Savory, "Digital filters for coherent optical receivers," *Optics Express*, vol. 16, no. 2, pp. 804–817, 2008.
13. M. Kuschnerov et al., "Energy efficient digital signal processing," Proceedings of Optical Fiber Communications, Th3E.7, 2014.
14. J. S. Wey et al., "Physical layer aspects of NG-PON2 standards Part 1: Optical link design [Invited]," *IEEE/OSA Journal of Optical Communications and Networking*, vol. 8, no. 1, pp. 33–42, 2016.
15. J. Zhang et al., "Real-time FPGA demonstration of PAM-4 burst-mode all-digital clock and data recovery for single wavelength 50G PON application," Proceedings of Optical Fiber Communications, M1B.7, 2018.
16. A. Eghbali et al., "Optimal least-squares FIR digital filters for compensation of chromatic dispersion in digital coherent optical receivers," *Journal of Lightwave Technology*, vol. 32, no. 8, pp. 1449–1456, 2014.
17. S. Kim et al., "Coherent receiver DSP implemented on a general-purpose server for a full software-defined access system," *IEEE/OSA Journal of Optical Communications and Networking*, vol. 11, no. 1, pp. A96–A102, 2019.
18. D. Lavery et al., "Reduced complexity equalization for coherent long-reach passive optical networks [invited]," *IEEE/OSA Journal of Optical Communications and Networking*, vol. 7, no. 1, pp. A16–A27, 2015.

7 Access Environment Considerations for Coherent Optics Systems

Zhensheng Jia and Jing Wang

CONTENTS

7.1 INTRODUCTION

When operators or service providers look to deploy coherent optics into their access networks, they are typically faced with two options: deploy coherent optics on the existing 10G system or build a new coherent-only connection. The ideal network for deploying such coherent systems would be a greenfield deployment on fibers without any compensation devices, such as dispersion compensation modules and other wavelength channels. However, in practice, to make the upgrade cost-effective, only one or a few channels may be considered in many brown field installations, depending on capacity demand, which means many of these networks that are deployed already with wavelength-division multiplexing (WDM) analog technology and/or 10G on–off keying (OOK) services will coexist with a coherent system to support a hybrid scenario over the same fiber transmission. Such a hybrid configuration needs to be investigated, especially the cross-phase modulation (XPM) impairment in the fiber nonlinear regime, to provide this option for operators to effectively support 100G on their existing networks.

Additionally, according to the information given by cable operators in a recent survey, 20% of existing cable access networks use a single-fiber topology. This means that downstream and upstream transmission to nodes takes place on a single strand of fiber. This number is expected to grow further in the near future. Therefore, bidirectional transmission is needed for coherent signals to support single-fiber topology and to facilitate the business use and redundancy of optical links. Full Duplex Point-to-Point (P2P) Coherent Optics uses the same wavelength in two directions over the same fiber at the same time. This technology doubles the fiber's capacity, significantly reduces the cost, and can be easily implemented in single-fiber networks without using any new chips for digital signal processing [1].

7.2   COEXISTENCE WITH LEGACY OPTICAL CHANNELS

The commercial coherent 100G transmission systems are showing excellent receiver sensitivity, robustness, and tolerance for channel impairments, such as chromatic dispersion (CD) and polarization mode dispersion (PMD). Therefore, the ideal network for deploying such coherent systems would be a greenfield deployment on fibers without any compensation devices, which is called a coherent-only implementation. However, in practice, there are many brown field installations, meaning many of these networks are deployed and have several WDM analog and/or 10G OOK (Ethernet over fiber or passive optical network) services running over the existing fiber already [2–6]. The expectation from the operators has been that adding additional 100G coherent services by using free channels in the WDM grid is preferred without impacting the existing services. This will essentially create a hybrid 10/100G network with multiple services coexistence. But the fact is that 10G signals based on analog amplitude modulation or OOK have a much higher power density than coherent 100G, causing them to have a much greater impact on the refractive index for nonlinear effects, such as XPM and four-wave mixing. Additionally, crosstalk penalties in International Telecommunication Union (ITU)-T grid networks with mixed rates lead to system degradation due to optical multiplexers/demultiplexers (Mux/Demux) in-band residual power or non-uniform channel grid allocation in dense WDM (DWDM) systems.

Three different kinds of optical multiplexers/demultiplexers have been evaluated: 8-port thin-film filters (TFFs), and 40- and 48-port array waveguide grating (AWG)–based optical Mux/Demux. TFFs use concatenated interference filters, each of which is fabricated with a different set of dielectric coatings designed to pass a single wavelength. TFFs have a better optical performance in terms of flatter passband ripple and higher isolation in neighboring channels. They work well for low channel counts, especially for analog WDM systems, but have challenges at higher channel counts and narrower spacing because they need several hundred layers of coating, which requires stricter error control. In contrast to TFFs, AWG devices use a parallel multiplexing approach that is based on planar waveguide technology. The key advantage of AWGs over TFFs is that their cost is not dependent on wavelength count making them extremely cost-effective for high-channel-count applications. The existing long-haul coherent DWDM systems are typically using AWG for Mux and Demux. In the testing, the insertion loss is ~1.5 dB for TFFs and ~3.5 dB for 40- and 48-port AWGs.

The coexisting setup was established with eight analog channels, two coherent 100G polarization multiplexing quadrature phase-shift-keying (PM-QPSK) channels (C-form factor pluggable 2 digital coherent optics form factor), two coherent 400G channels, and two 10G non-return zero (NRZ) channels. This coexistence hybrid scheme includes all the major modulation formats and services under different data rates/baud rates. A pair of 16-channel TFFT-based wavelength division multiplexers are used for channel multiplexing and demultiplexing.

The corresponding input power level is based on typical operational conditions for different detection schemes. The power levels are measured at the output port of each transmitter before entering the WDM Mux. Among them, the power of the analog channels is set to around 9.5 dBm; 56 GBaud 400G coherent channels have the power set to ~3 dBm; and 100G and 10G channels have the power between −1 and 0 dBm. To improve the spectral efficiency and confine the optical power within each WDM channel, the coherent signals are shaped by square-root-raised-cosine filters.

In summary, three main observations were found in the coexistence measurement experiments:

- Both coherent and analog/NRZ signals work well in the coexistence application with ~0.6-dB maximum power penalty in the case of 100-GHz channel spacing and 50/80-km fiber transmission distances for different nonlinearity tolerance scenarios.
- The legacy components/devices for analog systems are working well for coherent signals multiplexing and amplification, including analog erbium-doped fiber amplifier (EDFA), optical Mux, and Demux. Coherent signals show strong robustness when they are deployed in traditional analog DWDM systems.
- However, the conventional AWG-based optical Mux and Demux configuration, which is typically used for coherent channels, is not good for conventional analog channels.

7.3 FULL-DUPLEX COHERENT OPTICS

Today, achieving bidirectional transmission in an optical domain with a single laser requires two fibers. This is the standard practice using today's coherent optical technology. One laser in a transceiver performs two functional roles:

- as the optical signal source in the transmitter
- as the reference local oscillator signal in the receiver

Because of the use of the same wavelength from the same laser, a second fiber must be available for the other direction—one fiber for downstream and a second for upstream [7–9].

The second typical approach is to use a single fiber but transmit at different frequencies or wavelengths, similar to the upstream and downstream radio frequency (RF) spectrum split that is implemented in hybrid fiber-coaxial (HFC) networks. To accomplish this frequency/wavelength multiplexing approach, two lasers operating at different wavelengths are needed. Wavelength multiplexers and demultiplexers following a wavelength management and allocation strategy are needed to combine these different wavelengths over the same fiber. The burden of a second laser transcends economic concerns—increasing power consumption, operational complexity, and transceiver footprint.

To achieve the objective of keeping the cost down while using a single fiber, an alternative method is to employ full-duplex coherent optics. In this approach, an

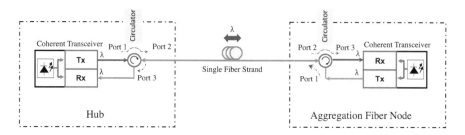

FIGURE 7.1   Full-duplex single-fiber approach (single wavelength).

optical circulator is added to each end of the coherent link in a special configuration. The circulator is a low-cost, passive, directional device that permits the coherent transceiver to transmit and receive using the same wavelength (and therefore using a single laser) over a single fiber in both directions simultaneously. Figure 7.1 shows one embodiment of this method with the circulator outside of the coherent transceiver for single wavelength connection in an unamplified single-fiber system.

Full-duplex coherent optics can employ WDM to support operation with other transceivers simultaneously. It is noted that the circulator or other optical directional elements could also be integrated with coherent transceiver, in which each wavelength will need an independent directional element before optical mux or after optical Demux.

A variety of unique aspects of access networks, taken together, enables the use of this approach. Unlike backbone and metropolitan coherent optical networks, access networks don't require multiple directional optical amplifiers in a cascade optical link. When comparing with intensity-modulation and direct-detection (IM-DD) system, coherent signals have a much higher Optical Signal-to-Noise Ratio (OSNR) sensitivity and tolerance to impairments from the spontaneous Rayleigh backscattering (continuous reflection) and Fresnel reflection (discrete reflections) than the intensity-modulated systems. In addition, the majority of existing optical systems in cable employ angle-polished connectors that provide excellent mitigation for return loss from Multiple-Path Interference, jumper cable/optical distribution panels/fusion, or mechanical splices. In addition, the threshold of the stimulated Brillouin scattering nonlinear effect is much suppressed because of the nature of phase-modulated signals on the reduction of optical carrier power and the increase of effective linewidth.

With this new dimension of direction-division multiplexing in the optical domain, any coherent wavelength can be used twice, once in each direction, thus doubling the whole fiber system capacity. This full-duplex implementation is not wavelength-selective. It works for both short and long wavelengths, and it can cover not only the entire C-band but also the entire fiber spectrum with different optical sources.

In full-duplex coherent optics, one of the critical components in the system is the optical circulator that enables the simultaneous bidirectional transmissions of the two optical paths without interference. It is based on Faraday magneto-optical effect, where through interactions between light and a magnetic field, the reciprocity of the optical transmission is destroyed and the light coming from the Port 1 will be guided to Port 2, while the light entering Port 2 will be redirected to Port 3. Thus, when connecting Port 1 to the optical transmitter, Port 2 to the long-distance fiber, and

Port 3 to the optical receiver, a transceiver to be used in full-duplex optical transmission system is built up. Traditional circulators are sensitive to temperature change, especially at the low-temperature conditions. Recently, thanks to the development of athermal circulators, thermal stability has been greatly improved. With humidity kept at 30%, it has been experimentally verified that stable isolations and insertion losses are achieved for an athermal circulator working from −40°C to +85°C.

7.4 DATA SECURITY IN OPTICAL FIBERS

In recent years, the ever-increasing cybercrime and the high costs paid by enterprises for user data breach have made network security a major concern in modern telecommunication, especially for data-sensitive and/or mission-critical applications for defense, government, healthcare, utility, and finance scenarios [10]. There are three data-protection measures: database security to protect active data being processed in data centers or servers; at-rest security to protect inactive data stored in data centers by controlling the user access and credentials; and "in-flight" security to protect data traveling in the network beyond the walls of data centers [11,12]. Both database and at-rest securities have been investigated extensively, and "in-flight" security deserves more attention since the privacy of data is most vulnerable during the data transmission when the security measures at data centers are no longer available.

Optical fiber used to be considered more secure than coaxial cables or wireless radio due to the absence of radiation and confinement of optical signals within a waveguide [13,14]. However, since most optical fibers are unguarded and can be easily tapped, data breaching by fiber tapping has become unpreventable. Since physical protection of optical fibers is impractical, instead "in-flight" encryption has become a requirement for most optical transport equipment [15–19].

Currently, there are two variants of encryption – unauthenticated and authenticated. Unauthenticated encryption does not increase the frame size and introduces no overhead but only provides data confidentiality [12]. Authenticated encryption, on the other hand, increases the frame size but provides enhanced security, including not only data confidentiality but also authentication, integrity protection, and replay protection [12]. The de facto standard for authenticated encryption is advanced encryption standard (AES) with Galois counter mode (GCM), where AES provides encryption, that is, data confidentiality, and GCM provides intrusion detection, intrusion prevention, and firewalling [11,12].

Network encryption can be done in different network layers, for example, optical transport network (OTN) encryption in layer 1 (physical layer (PHY)), media access control (MAC) security (MACSec) or Ethernet encryption at layer 2 (data link, MAC, Ethernet); Internet protocol (IP) security (IPSec) at layer 3 (network/IP); and secure sockets layer (SSL) and transport layer security (TLS) at layer 4 and above (transport, presentation and application layers).

Lower layer encryption offers simpler processing in terms of smaller overhead and latency, more efficient utilization of network bandwidth, and more comprehensive protection of higher layer protocols but has the penalty of reduced network compatibility. However, encryption at higher layers suffers from less efficient processing, larger overhead, and latency but provides finer encryption granularity, more flexibility, and improved network compatibility. Since over 99.9% cyberattacks target layers 3–7, it is common to encrypt at layer 1 (PHY) or 2 (Ethernet/data link) for optimal combination of efficiency, flexibility, and compatibility. In this session, we will focus on the layer 1 encryption for optical fibers. Other popular encryption solutions, such as MACSec, Ethernet encryption, and IPSec, are out of the scope of this book.

Based on ITU G.709 standard, OTN has become the de facto standard for 100G+ metro and core networks, supporting multiplexing, transport, and switching of all client types and protocols. Thanks to the versatility of OTN, layer 1 encryption is agnostic to protocol, client type, frame size, or data rate and can support services, including Ethernet, fiber channel and Synchronous optical networking/synchronous digital hierarchy (SONET/SDH) [15,16,19,20].

Many OTN vendors, such as Ciena, Infinera, and Microsemi, have implemented layer 1 encryption to protect the OTN frame payload and integrated it into their network transport equipment [15–20]. Low-cost solutions use unauthenticated encryption for data confidentiality only, whereas high-end solutions use authenticated encryption for authentication, integrity protection, and replay protection. Since authentication tags are carried by reserved bytes in the overhead of OTN frames, there is no padding or extension necessary to OTN frames. Thanks to the absence of overhead, layer 1 encryption can work at wirespeed, utilizing 100% throughput of the optical link. In theory, no real security could be achieved without frame overhead, since the two encryptors always have to exchange keys, but layer 1 encryption does have much smaller overhead and higher bandwidth utilization efficiency than any other encryption solutions at higher layers, which enables it to layer 2, MACSec extends the frame by up to 40 bytes and Ethernet encryption adds an overhead of 24–50 bytes per frame; at layer 3, IPsec adds up to 60% packet size expansion. On the other hand, limited by the small number of reserved bytes in OTN frames, the authentication capability offered by OTN encryption is very limited compared with higher layer encryption.

Most of the layer 1 encryption is implemented by an application-specific integrated circuit (ASIC) integrated with the network-transport equipment. Low latency as small as 180 ns can be achieved, which is much lower than layer 2 or 3 encryption, leaving plenty of margin for end-to-end latency budget. Due to its high throughput and low latency, layer 1 encryption is preferred for data-intensive and time-sensitive applications.

One disadvantage of layer 1 encryption is the lack of fine encryption granularity, which only supports bulk encryption per fiber or wavelength, making them only suitable for hop-by-hop instead of end-to-end applications. Different from end-to-end encryption, hop-by-hop encryption cannot support to network switching/routing without decryption, that is, if there is a node between the sending and receiving encryptors, the encrypted data has to be decrypted, routed or switched, and re-encrypted before sent out. Therefore, most layer 1 encryption solutions are restricted to point-to-point topologies with dedicated fiber links.

In recent years, due to the progress of layer 1 switching and the emergence of 100G metro switching network and packet-optical transport systems (P-OTS), finer encryption granularity, e.g., sub-wavelength, has become available at layer 1. Network traffic can be encrypted from 1.25 to 100 Gb/s, before or after multiplexing into a higher rate OTN signal, making end-to-end security also possible. Another side benefit of sub-wavelength encryption is to lower the entrance barrier of network encryption, so that smaller network operators can also afford the security service in a "pay-as-you-need" way.

On the current market, most commercial solutions for layer 1 encryption come from the manufacturers of network transport equipment, such as Infinera, Ciena, and Microsemi. All these solutions use separate sets of keys for authentication and encryption. For encryption, AES with 256-bit key size has become the de facto standard with the support of fast key rotation within seconds. For authentication, Infinera and Ciena use public key infrastructure (PKI) with X.509 certificate, whereas Microsemi uses Galois message authentication code for authentication. All these solutions claim to be compliant with US Federal Information Processing Standard (FIPS)-197. One feature that distinguishes Ciena from the others is the software-programmable flexible data rates from 100 to 400 Gb/s with 50 Gb/s increment.

All existing commercial solutions are based on vendor proprietary implementations, and no interoperability between different vendors is expected. Most layer 1 encryptors are implemented by an ASIC integrated in the transport equipment, which, on the one hand, makes them simple, low-cost, easy to operate, and saves precious space and power in data centers; on the other, it also creates the issue of double vendor lock-in. Customers have to purchase both network transport and encryption equipment from the same vendor, and most encryption functions are mandatory and there is no way to turn them off.

7.5   ENVIRONMENTAL CONDITIONS

A unique characteristic of the access environment is that many of its optical components will reside in an outdoor location subject to severe humidity and large temperature variation. In the case of the HFC networks, coherent transceivers may reside in aluminum clamshell enclosures that are network powered. In the case of telco and other cable scenarios, coherent transceivers could also reside in outdoor cabinets. It is expected that coherent optics transceivers and other optical components supporting coherent optics transport will have to support startup temperatures as low as −40°C to internal temperatures as high as +85°C. Unlike long-haul and metro scenarios where optical transmitters and receivers are always in temperature control environments, in the access, there are many scenarios with harsh environmental requirements.

REFERENCES

1. Z. Jia, L. A. Campos, M. Xu, H. Zhang, J. Wang, C. Knittle, "Impact of access environment in cable's digital coherent system – coexistence and full duplex coherent optics," SCTE/ISBE Cable-Tec Expo'18, Oct. 2018.
2. C. Liu, L. Zhang, M. Zhu, J. Wang, L. Cheng and G. Chang, "A novel multi-service small-cell cloud radio access network for mobile backhaul and computing based on radio-over-fiber technologies," *Journal of Lightwave Technology*, vol. 31, no. 17, pp. 2869–2875, 2013.
3. J. Nakagawa, M. Noda, N. Suzuki, S. Yoshima, K. Nakura and M. Nogami, "Demonstration of 10G-EPON and GE-PON coexisting system employing dual-rate burst-mode 3R transceiver," *IEEE Photonics Technology Letters*, vol. 22, no. 24, pp. 1841–1843, 2010.

4. B. Zhu, D. Au, F. Khan and Y. Li, "Coexistence of 10G-PON and GPON reach extension to 50-km with entirely passive fiber plant," Proceedings of European Conference on Optical Communication, pp. 1–3, 2011.

5. H. Rohde, E. Gottwald, Sönke Rosner, Erik Weis, Paul Wagner, Yuriy Babenko, Daniel Fritzsche, and Hacene Chaouch, "Field trials of a coherent UDWDM PON: Real-time LTE backhauling, legacy and 100G coexistence," Proceedings of European Conference on Optical Communication, pp. 1–3, 2014.

6. X. Hu, C. Ye, and K. Zhang, "Converged mobile fronthaul and passive optical network based on hybrid analog-digital transmission scheme," Proceedings of Optical Fiber Communications, 2016, paper W3C.5.

7. W. Lee, M. Y. Park, S. H. Cho, J. Lee, C. Kim, G. Jeong, and B. W. Kim, "Bidirectional WDM-PON based on gain-saturated reflective semiconductor optical amplifiers," *IEEE Photonics Technology Letters*, vol. 17, no. 11, pp. 2460–2462, 2005.

8. E. S. Son, K. H. Han, J. K. Kim and Y. C. Chung, "Bidirectional WDM passive optical network for simultaneous transmission of data and digital broadcast video service," *Journal of Lightwave Technology*, vol. 21, no. 8, pp. 1723–1727, 2003.

9. M. G. Larrode, A. M. J. Koonen, J. J. V. Olmos and A. Ng'Oma, "Bidirectional radio-over-fiber link employing optical frequency multiplication," *IEEE Photonics Technology Letters*, vol. 18, no. 1, pp. 241–243, 2006.

10. C. Jaggi, "Evaluation guide: layer 2 encryptors for metro and carrier Ethernet," whitepaper of inside-it.ch, version 1.3, Jan. 2015. www.uebermeister.com/files/inside-it/2014_Evaluation_Guide_Encryptors_Carrier_and_Metro_Ethernet.pdf

11. C. Jaggi, "Ethernet encryptors for metro and carrier Ethernet, an introduction," whitepaper of inside-it.ch, version 6.04, Jun. 2017. www.uebermeister.com/files/inside-it/2016_Introduction_Encryption_Metro_and_Carrier_Ethernet.pdf

12. C. Jaggi, "Market overview Ethernet encryptors for carrier Ethernet, MPLS and IP networks," whitepaper of inside-it.ch, version 6.1, Jun. 2017. www.uebermeister.com/files/inside-it/2017_market_overview_Ethernet_encryptors_for_Metro_and_Carrier_Ethernet.pdf

13. ID Quantique whitepaper, "Fibre optic networks: your weakest link?" version 1.0, Mar. 2011. https://marketing.idquantique.com/acton/attachment/11868/f-0209/1/-/-/-/-/Optical%20Fibre%20Vulnerability_White%20Paper.pdf

14. ID Quantique whitepaper, "State-of-the-art network encryption architecture & best practices," version 2, Aug. 2017. https://marketing.idquantique.com/acton/attachment/11868/f-020c/1/-/-/-/-/State-of-the-Art%20Network%20Encryption_White%20Paper.pdf Ciena

15. Ciena whitepaper, "Carrier-grade Ethernet as a foundation for assured data networks," www.ciena.com/insights/white-papers/393551021.html

16. Ciena application note, "Wavelogic encryption solutions, securing all in-flight data, all the time," https://media.ciena.com/documents/Wavelogic_Encryption_Solution_AN.pdf

17. Ciena data sheet, "High-capacity wire-speed encryption modules for the 6500 packet-optical platform," https://media.ciena.com/documents/High_Capacity_Wire_Speed_Encryption_Modules_DS.pdf

18. Ciena whitepaper, "Encryption testing and certification," https://media.ciena.com/documents/Encryption_testing_and_certification_PS.pdf

19. J. Gill, "Enhancing data center security with in-flight encryption," Infinera presentation at INTEROP 17. www.infinera.com/wp-content/uploads/Infinera-Interop-2017-Future-of-Networking-Summit.pdf

20. Microsemi poster, "OTN 3.0 enabling 100G beyond optical transport," www.microsemi.com/document-portal/doc_download/1243223-otn-3-0-poster

8 Photonics Integrated Circuits for Coherent Optics

Haipeng Zhang

CONTENTS

8.1 INTRODUCTION

Wired optical access technologies of future will have to take full advantages of evolutions in the field of optics and photonics to support the exponential growth in demand for bandwidth. In the last decades, great progresses have been made to the integration of optical components on a single chip, also known as photonic integrated circuit (PIC). As of today, various material platforms have been adopted for photonics integration, each with their unique advantages as well as shortcomings. Among them, Silicon on Insulator (SOI) and III–V semiconductors (such as Indium-Phosphide (InP)) representing material platform workhorses deliver high-performance dense integration solutions fueling today's high-bandwidth optical interconnects. This chapter reviews the state-of-the-art technologies and performance on the basis of the two material platforms and development roadmaps for applications in next-generation optical access networks.

8.2 SILICON PHOTONICS

Silicon Photonics (SiPh) benefits from the complementary metal–oxide–semiconductor (CMOS) technology platform that has dominated the microelectronics industry more than 40 years. SiPh-based PIC has become one of the most promising technologies to provide highly integrated and low-cost components and systems for future optical communication networks [1,2]. The vast potential of integrating complex devices and systems on a SOI platform was first recognized in the pioneer studies during the mid-1980s [1]. Over the last two decades, significant progresses have been made in the field of photonic integrated circuits. SiPh technology has reached new heights in terms of device performance and levels of integration. Monolithic integration of a wide range of components such as positive–intrinsic–negative (p-i-n) junction modulators, Germanium (Ge)- and SiGe-based photodetectors, polarization beam splitters/combiners, and other passive devices has been demonstrated with good performances. SiPh-based optical transceiver products from different companies such as Luxtera, Kotura (now Mellanox), and Acacia Communications have shipped out in significant volumes to support datacenter interconnect, telecommunications, and high-performance computing markets [2].

Compared with other integrated photonics platforms such as III–V semiconductors (InP), although SiPh excels in many aspects such as mature fabrication/testing processes and better yield control, one major challenge is the lack of efficient optical sources due to silicon's indirect bandgap structure. In general, today's SiPh technology heavily relies on utilizing III–V gain components to provide efficient electrically pumped optical sources. There are three approaches among many to achieve this on the basis of their potential to deliver high-performance integrated optical light source solutions. One approach is to have an external III–V laser coupled to the silicon chip that integrated the majority of optical functions. While this approach has many advantages such as good silicon PIC performance, separate yield control of silicon passive and light source, silicon PIC CMOS processing, etc., the major disadvantages are the extra cost and complexity associated with the separate light source testing and packaging processes. Despite having these drawbacks, commercial products based on this process have already been released by some companies such as Acacia, Luxtera, Cisco, and PETRA [3]. Another path towards light source integration on an SiPh platform is by bonding III–V chips on processed SOI wafer with coarse alignment, followed by III–V chip processing, also called heterogeneous integration. This process allows efficient wafer scale integration with significantly reduced tolerance requirements for III–V chip alignment. It is actively investigated by both academia (UCSB [4,5], Ghent University [6]), and industry (Intel [7], Juniper Networks [8], and HP Enterprise [9]) with products for data centers being shipped in volume [7]. The major drawback of this approach is the cost to implement chip-to-wafer bonding and the laser processing, which can affect the yield of the entire package. The third approach, an active research area in both academia and industry players (such as AIM Photonics, IMEC, and NTT) [3,10], is the direct epitaxial growth of III–V layers on silicon using buffer layers. This process potentially offers the lowest cost technology suit among others by providing a complete monolithic active–passive solution. Early stage studies in this field show limited light source lifetime and reliability due to defect-induced material degradation and thermal expansion mismatch between III–V and silicon [3]. Through a combination of using quantum dot active regions and careful defect engineering, the state-of-the-art evaluation shows a record lifetime of more than 10,000,000 h [11], making this approach a very promising solution for the monolithic integration of light sources on a next-generation SiPh platform.

Although integrating light sources on SiPh platforms generally requires extra efforts compared with III–V generic integration, an important benefit associated with this technology is the great potential of making widely tunable lasers of ultra-narrow linewidth. Low-phase noise light sources become increasingly important in modern coherent optical communications that typically utilize advanced modulation formats to achieve high data rate at a single wavelength. Traditional III–V lasers, which carry out both photon generation and photon storage in the same III–V material, suffer from insufficient quality factor (Q) due to free-carrier absorption. As comparison, SiPh with either heterogeneously/monolithically integrated or externally coupled III–V light sources benefits from utilizing high-Q photon storage in silicon and photon generation in III–V separately, which allows one to achieve sub-kHz linewidth [12].

Silicon optical modulators are also a key focus area of SiPh. Early research of silicon modulators can be traced back to the mid-1980s. Significant progress has been made recently with high-performance modulators demonstrated based on different technologies. Among many approaches, Mach–Zehnder modulator (MZM)–based devices that utilize the plasma dispersion effect in reverse-biased p–n/p–i–n junctions have been deployed in commercial products. As a mature technology, silicon MZM features athermal operation and wide optical bandwidth in the order of tens of nm. The state-of-the-art work reported modulation efficiencies in the order of 0.74 V·cm (half-wave voltage of 4.1 V) and a 3-dB electro-optic bandwidth of up to 48 GHz [13]. The disadvantages, however, are the relatively large drive voltage (usually >4 V) that is generally considered incompatible with a CMOS electronic driver, and the long device length (in the order of mm); both lead to a large power consumption. Another mature silicon modulator solution that has been commercialized is the SiGe-based electro-absorption modulator (EAM) [14]. Operation of this type of devices relies on electrically shifting the absorption edge of the SiGe material. Although this device requires relatively low driving voltage and has a compact size, it usually operates over a limited temperature and wavelength range.

Recently, great progress has been made in producing high-performance optical modulators on a variety of novel material platforms, such as Lithium-Niobate (LN) on silicon and polymer on silicon, etc. By bonding an LN thin film on a SOI substrate and creating LN waveguides by using carefully tuned dry etch process, good optical confinement can be achieved, thus great electro-optic efficiencies and performance [15]. Combining the conventional SOI waveguides with highly efficient organic electro-optic materials, silicon-organic-hybrid MZM with a π-voltage of 1.6 V to generate 100 GBd 16 quadrature amplitude modulation (QAM) signal has been reported [16]. These technologies provide alternative paths toward high-performance modulators with low drive voltage and high bandwidth on silicon, yet they haven't reached the level of maturity required for commercialization.

Photodetectors are also essential components for SiPh-based optical communication systems. Although silicon can be excellent materials for passive devices, it is rather a poor optical absorber in the near-infrared (NIR) wavelength range. Hybrid integration of III–V materials on silicon is less desired due to its complexity and high cost [2]. Ge, which has a strong absorption in the NIR, is considered an ideal material for optical detection. Over the last decades, great effort has been made to overcome the major challenge of Ge on silicon lattice mismatch that prevented the monolithic integration of Ge detectors on silicon in the past. Today, Ge-on-Si photodetectors are among the most mature devices on SiPh platforms, offering good performances such as >40 GHz bandwidth, around 1 A/W response, and <1 μA dark current [17,18], which is directly comparable to their III–V counterparts.

Overall, an SiPh platform is an emerging area of focus for high-speed optical communication systems. Evolving research and manufacturing technologies today allow creating highly integrated and high-performance optical components on a single silicon chip for compact optical modules in C-form factor pluggable 2 (CFP2) or quad small form-factor pluggable (QSFP) for next-generation optical access networks. As a challenger of today's well-established discrete-component technologies based on III–V or LN, SiPh has strong potential of evolving into low-cost system-on-chip solutions like today's CMOS-based micro/nanoelectronics.

8.3 INP PHOTONICS

InP is another material platform that offers the state-of-the-art performance for optoelectronic devices operating in the 1,300–1,600-nm wavelength window [19]. As a direct bandgap III–V compound semiconductor material, InP photonics integrated circuit enables the monolithic integration of laser sources with other components such as modulators, amplifiers, multiplexers, and detectors in a wafer-scale processes. Active and passive interfaces are joined via vertical twin or single guide growth, quantum well intermixing, and/or butt-joint regrowth [20]. As a firmly established solution, InP PIC-based optical transceivers accounted for a 1-B$ market share in 2015 and are expected to grow beyond 3 B$ in 2020 [21]. Partnered with Oclaro, Fraunhofer HHI, SMART Photonics, and LioniX, the joint European platform for InP-based PIC – JePPIX offers open-access InP PIC foundry services. Companies such as Finisar, Infinera, and Lumentum offer commercially available products today based on InP PICs.

By leveraging the monolithic integration capability of InP photonics, the fabrication of PIC light sources such as distributed feedback (DFB) and distributed Bragg reflector (DBR) lasers becomes a very mature process. In optical transceivers that feature intensity-modulation and direct-detection (IM-DD) formats, high-speed directly modulated lasers are highly desirable as they can significantly reduce the cost and additional loss incurred by a separate optical modulator. A direct modulation of III–V DFB laser up to 40 Gb/s for a device length of 100 μm has been demonstrated [22]. Later, a DFB laser with a low-threshold current of 5.6 mA at 85°C has been used to demonstrate 50 Gb/s data transfer over 10 km fiber [23]. As the insatiable demand for data-intensive applications and services continues, IM-DD solutions are reaching a saturation point. To support even a higher data rate, coherent technology, which encodes data not only in amplitude but also phase and polarization of light, is now receiving growing attention. In a coherent system, high-quality light sources with low-phase noise and thus narrow spectral linewidth are essential. On an InP platform, DFB lasers with a narrow linewidth of <160 kHz and a wide wavelength tuning range of 40 nm have been demonstrated, with an achieved output power of >13 dB m by integrating with a semiconductor optical amplifier (SOA) [24]. More recently, a thermally tuned DBR laser with SOA that is fully integrated on an InP platform achieved 70 kHz spectral linewidth and 17 dB m output power over a 41-nm wavelength tuning range [25]. These narrow-linewidth lasers, monolithically combined with high-performance InP modulators in a dense wavelength division multiplexing (DWDM) system, ensure a scalable solution to meet the ever-rising demands for a higher data rate.

As a material with strong electro-optic effect, InP is also ideal to make modulators that feature high modulation bandwidth, low drive voltage, and compact size. Early generation InP-based PICs utilized EAM combined with DFB or DBR lasers [26] with relatively limited operating wavelength bandwidth. Later, MZMs are adopted and integrated with lasers and SOAs to achieve an integrated transmitter PIC that is capable of tuning over a much wider wavelength range. Most recently, coherent technology has gained increasing attention due to its superior performance and vast potential. As a result, nested interferometric modulators on InP platforms that can handle complex modulation formats in phase, amplitude, and polarization have been widely adopted. In terms of device's performance, an InP-based in-phase/quadrature (I/Q) modulator with a 3 dB electro-optic bandwidth of 40 GHz and a half-wave voltage of 2 V has been demonstrated [27]. Similar results were published the same year with a lower half-wave voltage of 1.5 V (voltage-length product of 0.6 V cm) [28]. Later, an InP-based I/Q modulator with a record high bandwidth of 67 GHz and a half-wave voltage of 1.5 V (voltage-length product of 0.54 V cm) was achieved using a novel n–i–p–n heterostructure [29].

Benefited from $In_xGa_{1-x}As$ alloy, a variable band gap III–V semiconductor that can be monolithically integrated, an InP-based optical receiver is one of the most mature systems at NIR wavelengths. The classic photodetector type is the p–i–n structure that relies on both carrier drift and diffusion to generate photocurrent. Great progresses have been made in the past few years, especially on balanced p–i–n photodetectors for high-speed coherent optical communication systems. A balance p–i–n photodetector with a 3-dB bandwidth of 80 GHz and a responsivity of 0.55 A/W was demonstrated earlier [30]. Later, the 3-dB bandwidth was pushed further to 90 GHz [31]. More recently, a balance p–i–n photodetector with a record high bandwidth of 115 GHz was demonstrated on an InP substrate [32]. Avalanche photodiodes (APDs), which operate at high electrical fields to achieve photocurrent gain of >1, offer improved sensitivity compared to the p–i–n photodetectors. By using narrow InGaAs multiplication layer width, InP-based APD with a low dark current of 10–300 nA with a 3-dB bandwidth of 80 GHz was achieved [33].

8.4   SUMMARY

In this chapter, we reviewed the state-of-the-art technologies in the field of photonics integrated circuits. Photonics integration on an InP platform is a mature technology today and the workhorse of performance focused market segments such as long-haul and coherent communications. Compared with SiPh-based PICs, InP excels in performance of active components such as modulators and photodetectors, in terms of their modulation bandwidth, power consumption, and noise figure. In particular, monolithic laser integration with the rest of the PIC on an InP platform is one of the major advantages over SiPh, typically requiring advanced integration/packaging technologies for light sources. However, InP PICs do suffer from a very small wafer size of up to 100 mm, limiting the level of integration and the total number of devices on a single wafer, which potentially leads to lower overall yield and higher cost. On the other hand, by leveraging today's highly mature CMOS processing/testing capabilities, SiPh-based PIC is fabricated on 300-mm wafers with very high yield, allowing for the integration of a large number of devices with unprecedented complexity. Furthermore, co-package Si-based electronics with InP-based PICs has proven to be more challenging compared to SiPh. Overall, the two technology platforms will coexist in the near future and remain critical with their own technology roadmap directed at improving overall product yield and reduce cost for optical links.

REFERENCES

1. A. E. Lim et al., "Review of silicon photonics foundry efforts," *IEEE Journal of Selected Topics in Quantum Electronics*, vol. 20, no. 4, pp. 405–416, 2014.
2. D. Thomson et al., "Roadmap on silicon photonics," *Journal of Optics*, vol. 18, no. 7, pp. 1–20, 2016.
3. J. E. Bowers and A. Y. Liu, "A comparison of four approaches to photonic integration," 2017 Optical Fiber Communications Conference and Exhibition (OFC), Los Angeles, CA, pp. 1–3, 2017.
4. T. Komljenovic, et al., "Photonic integrated circuits using heterogeneous integration on silicon," *Proceedings of the IEEE*, vol. 106, no. 12, pp. 2246–2257, 2018.
5. T. Komljenovic et al., "Heterogeneous silicon photonic integrated circuits," *Journal of Lightwave Technology*, vol. 34, no. 1, pp. 20–35, 2016.
6. G. Roelkens et al., "III–V-on-silicon photonic devices for optical communication and sensing," *Photonics*, vol. 2, no. 3, pp. 969–1004, 2015.
7. Intel. Accessed: Jul. 7, 2017. www.intel.com/content/www/us/en/architectureand-technology/silicon-photonics/opticaltransceiver-100g-psm4-qsfp28-brief.html
8. B. R. Koch et al., "Integrated silicon photonic laser sources for telecom and datacom," Optical Fiber Communication Conference/National Fiber Optic Engineers Conference, p. 8, 2013. Paper PDP5C.
9. D. Liang, et al., "Integrated finely tunable microring laser on silicon," *Nature Photonics*, vol. 10, no. 11, pp. 719–722, 2016.
10. A. Y. Liu and J. Bowers, "Photonic integration with epitaxial III–V on silicon," *IEEE Journal of Selected Topics in Quantum Electronics*, vol. 24, no. 6, pp. 1–12, 2018.
11. D. Jung et al., "Impact of threading dislocation density on the lifetime of InAs quantum dot lasers on Si," *Applied Physics Letters*, vol. 112, no. 15, p. 153507, 2018.
12. C. T. Santis et al., "High-coherence semiconductor lasers based on integral high-Q resonators in heterogeneous Si/III -V platforms," *Proceedings of the National Academy of Sciences*, vol. 111, pp. 2879–2884, 2014.
13. S. S. Azadeh, et al., "Low V silicon photonics modulators with highly linear epitaxially grown phase shifters". *Optics Express*, vol. 23, pp. 23526–23550, 2015.
14. D. Feng et al., "High speed GeSi electro-absorption modulator at 1550 nm wavelength on SOI waveguide", *Optics Express*, vol. 20, pp. 22224–22232, 2012.
15. C. Wang, et al., "Integrated lithium niobate electro-optic modulators operating at CMOS-compatible voltages", *Nature*, vol. 562, pp. 101–104, 2018.
16. S. Wolf, et al., "Coherent modulation up to 100 GBd 16QAM using silicon-organic hybrid (SOH) devices", *Optics Express*, vol. 26, pp. 220–232, 2018.
17. L. Vivien, et al., "Zero-bias 40Gbit/s germanium waveguide photodetector on silicon", *Optics Express*, vol. 20, pp. 1096–1101, 2012.
18. C. T. DeRose et al., "Ultra compact 45 GHz CMOS compatible Germanium waveguide photodiode with low dark current," *Optics Express*, vol. 19, pp. 24897–24904, 2011.
19. K. Williams, "InP integrated photonics: State of the art and future directions," 2017 Optical Fiber Communications Conference and Exhibition (OFC), Los Angeles, CA, pp. 1–3, 2017.

20. M. Smit et al., "An introduction to InP-based generic integration technology," *Semiconductor Science and Technology*, vol. 29, 2014. DOI:10.1088/0268-1242/29/8/083001

21. JePPIX ROADMAP 2018.

22. W. Kobayashi, et al., "40-Gbps direct modulation of 1.3-μm InGaAlAs DFB laser in compact TO-CAN package," Optical Fiber Communication Conference/National Fiber Optic Engineers Conference 2011, OSA Technical Digest (CD) (Optical Society of America, 2011), 2011. Paper OWD2.

23. K. Nakahara et al., "Direct modulation at 56 and 50 Gb/s of 1.3-μm InGaAlAs ridge-shaped-BH DFB lasers," *IEEE Photonics Technology Letters*, vol. 27, no. 5, pp. 534–536, 2015.

24. H. Ishii, et al., "Narrow spectral linewidth operation (≪160 khz) in widely tunable distributed feedback laser array," *Electronics Letters*, vol. 46, no. 10, pp. 714–715, 2010.

25. M. C. Larson et al., "Narrow linewidth sampled-grating distributed Bragg reflector laser with enhanced side-mode suppression," 2015 Optical Fiber Communications Conference and Exhibition (OFC), Los Angeles, CA, pp. 1–3, 2015.

26. F. Kish et al., "System-on-chip photonic Integrated Circuits," *IEEE Journal of Selected Topics in Quantum Electronics*, vol. 24, no. 1, pp. 1–20, 2018.

27. E. Rouvalis, "Indium phosphide based IQ-modulators for coherent pluggable optical transceivers," IEEE Compound Semiconductor Integrated Circuit Symposium, New Orleans, LA, pp. 1–4, 2015.

28. G. Letal et al., "Low loss InP C-band IQ modulator with 40GHz bandwidth and 1.5V Vπ," 2015 Optical Fiber Communications Conference and Exhibition (OFC), Los Angeles, CA, pp. 1–3, 2015.

29. Y. Ogiso, et al., "Over 67 GHz bandwidth and 1.5 V Vπ InP-based optical IQ modulator With n-i-p-n heterostructure," *Journal of Lightwave Technology*, 35, pp. 1450–1455, 2017.

30. P. Runge, et al., "80GHz balanced photodetector chip for next generation optical networks," Optical Fiber Communications, 2014. Paper M2G.3.

31. P. Runge, et al., "Polarisation Insensitive coherent receiver PIC for 100Gbaud communication," Optical Fiber Communications, 2016. Paper Tu2D.5.

32. P. Runge et al., "Waveguide integrated balanced photodetectors for coherent receivers," *IEEE Journal of Selected Topics in Quantum Electronics*, vol. 24, no. 2, pp. 1–7, 2018.

33. L. E. Tarof, "Planar InP-InGaAs avalanche photodetectors with n-multiplication layer exhibiting a very high gain-bandwidth product," *IEEE Photonics Technology Letters*, vol. 2, no. 9, pp. 643–646, 1990.

9 Standard Development

Zhensheng Jia and Curtis Knittle

CONTENTS

9.1 INTRODUCTION

In the past, standardization in the optical industry was driven mainly by short-reach metro/aggregation applications, where optical performance is not a differentiator. In long-distance transmission, the use of discrete performance-optimized optical components and best-in-class proprietary digital signal processing (DSP) algorithms was critically important to gain a competitive advantage in coherent optical 100G deployments. When the coherent optics extends its application into metro and access aggregation field, it is highly preferable that coherent transceivers from different system vendors be interoperable at the optical transport layer to simplify the deployment of multivendor networks. The standardization of coherent interconnects has been making significant progress in the last 2 years. The momentum behind these efforts continues to grow as it becomes clear that standardization of coherent technologies can be achieved. Today, around seven DSP solutions from different companies are commercially available. Standardization will eventually lead to improved interoperability and predictable performance and allow operators to utilize optical fiber infrastructure and more to efficiently meet future bandwidth demand.

9.2 OIF

Metro-access data center interconnections (DCI) are point-to-point (P2P) fiber connections with a reach of typically less than 120 km, well beyond the capabilities of direct-detection datacom transceivers. Conventional coherent optical transceivers designed for backbone applications can be used. They have, however, higher performance than required, are more complex, and are generally offered at a higher price point. These transceivers are not interoperable; they are based on proprietary solutions. In response to this, the Optical Internetworking Forum (OIF) launched a new project in 2016 related to coherent transmission technology: 400G ZR Interoperability. The OIF expects to develop an implementation agreement (IA) for

400G ZR and short-reach dense wavelength-division multiplexing (DWDM) multivendor interoperability. The IA will help to promote interoperability among coherent optical transceivers and transponders for router-to-router interconnect and other passive single-channel and amplified DWDM applications at distances up to 120 km. The IA will support single-carrier 400G transmission using coherent detection and advanced DSP/forward error correction (FEC) algorithms. The major application is in DCI using concatenated FEC (cFEC) with soft-decision inner Hamming code and hard-decision (HD) outer Staircase code. This cFEC provides a net coding gain (NCG) of 10.8 dB and pre-FEC bit error rate (BER) of 1.22E−2 for coherent dual-polarized 16-quadrature amplitude modulation (QAM) modulation format at a baud rate of ~60 GBaud and 15 W target power consumption. The potential form factors are high faceplate density quad small form-factor pluggable double density and Octal Small Form-Factor Pluggable. There is considerable investment underway to develop 400G ZR transceivers, and the first commercial release of these products is scheduled for the second half of 2020. 400G ZR+ is a new concept where vendors employ more powerful soft-decision FEC (SD-FEC) and higher performance coherent optics to improve performance and broaden the available applications. The specifications and standardization of the performance-enhanced 400G ZR+ are still evolving [1].

9.3   CABLELABS

CableLabs recognized the benefits of coherent optics in access networks and announced the launch of the P2P Coherent Optics specifications development activities in 2017. This allows the cable industry to support the growing capacity requirements of broadband access as it evolves toward distributed access architectures and fiber deeper topologies, and a substantial increase in volume of optical connections to intelligent nodes increases substantially. On June 29, 2018, CableLabs publicly unveiled for the first time two new specifications: P2P Coherent Optics Architecture Specification and P2P Coherent Optics Physical Layer v1.0 Specification. These two new specifications are the result of a focused effort by CableLabs, its members, and the manufacturer partners to develop coherent optics technology for the access network and to bring coherent optical technology to market quickly. On March 12, 2019 CableLabs announced another addition to its family of P2P Coherent Optics specifications: the P2P Coherent Optics Physical Layer v2.0 Specification. This new specification defines interoperable P2P Coherent Optics links running at 200 Gbps (200G) on a single wavelength [2].

9.4   IEEE, ITU-T, AND OPEN ROADM

The Institute of Electrical and Electronics Engineers (IEEE) continues to develop a set of standards to define the physical and data link layers of optical Ethernet, the latest activity being a short-reach coherent optical standard. IEEE P802.3ct (100 and 400 Gbps over DWDM systems) is to provide a physical layer specification supporting 100 and 400 Gbps operation on a single wavelength capable of at least 80 km over a DWDM system [3]. While the main drivers for this effort have been multi-system operators and DCI, it is easy to see how these solutions could be used for future mobile aggregation and core backhaul. The IEEE 802.3ct project will leverage the OIF 400 ZR specification, and it is scheduled to finish the standard in the fall of 2021.

International Telecommunication Union (ITU)-T Study Group 15 has also standardized coherent line interfaces at 100 Gbps data rate. The revised ITU standard – ITU G.698.2 "Amplified multichannel DWDM applications with single channel optical interfaces" – has added new 100G DWDM application codes using dual-polarization (DP)–differential quadrature phase-shift-keying (DQPSK) modulation format in two different application areas: first for operation over P2P WDM systems for reaching approximately 80 km (typical data-center-interconnect use case) where new codes have been specified, and second for operation over amplified metro Reconfigurable Optical Add/Drop Multiplexer (ROADM) networks for reaching up to about 450 km [4]. Work has been also initiated for a future revision of G.698.2 to specify 200G and 400G application codes, with DP-16 QAM provisionally agreed as the modulation format for 400G applications. A few amendments provide flexible interoperable short- and long-reach optical transport network (OTN) frame formats for operation over 100G, 200G, and 400G interfaces, including G.709.1: flexible OTN short-reach interfaces; ITU G.709.2 "OTU4 long-reach interface"; and ITU G.709.3 "Flexible OTN long-reach interfaces".

Another standardization effort from the optical industry is open ROADMs, which defines ROADM interoperability specifications. This includes the ROADM switch as well as transponders and pluggable optics. The consortium was initiated by AT&T, and now it has nine service providers and seven system vendors participating. Open ROADM has specified 100, 200, 300, and 400G transponder optical parameters for line-side interoperable capability [5].

9.5   FORWARD ERROR CORRECTION

Besides the framing and modulation format, FEC is another essential element that needs to be defined to enable the development of interoperable transceivers using optical technology. The industry trends are currently moving toward removing proprietary aspects and becoming interoperable when the operators advocate more open and disaggregated transport in high-volume short-reach applications [6].

When considering which FEC to choose for a new specification, you need to consider some key metrics, including the following:

- Coding overhead rate – The ratio of the number of redundant bits to information bits
- NCG – The improvement of received optical sensitivity with and without using FEC associated with increasing bit rate
- Pre-FEC BER threshold – A predefined threshold for error-free post–FEC transmission determined by NCG

Other considerations include hardware complexity, latency, and power consumption.

One major decision point for FEC coding and decoding is between HD-FEC and SD-FEC. HD-FEC performs decisions whether 1s or 0s have occurred based on exact thresholds, whereas SD-FEC makes decisions based on probabilities that a 1 or 0 has occurred. SD-FEC can provide higher NCG to get closer to the ideal Shannon limit with the sacrifice of higher complexity and more power consumption [7,8].

The first-generation FEC code, standardized for optical communication, is Reed–Solomon (RS) code. RS is used for long-haul optical transmission as defined by ITU-T G.709 and G.975 recommendations. In this RS implementation, each codeword contains 255 codeword bytes, of which 239 bytes are data and 16 bytes are parity, usually expressed as RS (255,239) with the name of Generic FEC. Several FEC coding schemes were recommended in ITU-T G. 975.1 for high bit-rate DWDM submarine systems in the second generation of FEC codes. The common mechanism for increased NCG was the use of concatenated coding schemes with iterative HD decoding. The most commonly implemented example is the Enhanced FEC from G.975.1 Clause I.4 for 10 and 40G optical interfaces.

At the 100-Gbps data rate, CableLabs has adopted HD Staircase FEC, defined in ITU-T G.709.2, and included in CableLabs P2P Coherent Optics Physical Layer v1.0 (PHYv1.0) Specification [2]. This Staircase FEC, also known as high-gain FEC, is the first coherent FEC that provides an NCG of 9.38 dB with the pre-FEC BER of 4.5E−3. The 100G line-side interoperability has been verified in the very first CableLabs' P2P Coherent Optics Interoperability Event.

At the 200-Gbps data rate, openFEC (oFEC) was selected in CableLabs most-recent release of P2P Coherent Optics PHYv2.0 Specification [2]. The oFEC provides an NCG of 11.1 dB for QPSK with pre-FEC BER of 2E−2 and 11.6 dB for 16-QAM format after three soft-decision iterations to cover multiple use cases. This oFEC was also standardized by Open ROADM–targeting metro applications [5].

Although CableLabs has not specified 400G coherent optical transport, the Optical Interworking Forum (OIF) has adopted a 400G cFEC with soft-decision inner Hamming code and HD outer Staircase code in its 400G ZR standard [1]; this same FEC has been selected as a baseline proposal in the IEEE 802.3ct Task Force. This 400G IA provides an NCG of 10.8 dB and pre-FEC BER of 1.22E−2 for coherent dual-polarized 16-QAM modulation format, especially for the DCI.

Table 9.1 summarizes performance metrics for standardized FEC in optical fiber transmission systems.

TABLE 9.1
Standardized FEC in Coherent Optical System

HD: Hard Decision SD: Soft Decision	Coding Overhead Rate	Net Coding Gain (NCG) dB @10E−15	pre-FEC BER Threshold
GFEC (HD)	6.69%	6.2	8.0E−5
EFEC (HD)	6.69%	8.67	2.17E−3
100G: Staircase FEC (HD)	6.69%	9.38	4.5E−3
200G: oFEC (SD)	15.3%	11.1 for QPSK 11.6 for 16-QAM	2.0E−2
400G: cFEC (SD)	14.8%	10.4 for QPSK 10.8 for 16-QAM	1.22E−2

9.6   SUMMARY

All these standardization activities reinforce the view of coherent optics moving to shorter reach and high-volume applications, such as in the access network. It is believed that the industry as a whole will benefit from successful standardization of coherent transceivers, including both optical performance and DSP functions and capabilities.

REFERENCES

1. 400ZR, www.oiforum.com/technical-work/hot-topics/400zr-2/
2. "P2P Coherent Optics Physical Layer 2.0 Specification – P2PCO-SP-PHYv2.0-I01-190311," 11 March 2019, Cable Television Laboratories, Inc.
3. www.ieee802.org/3/ct/public/index.html
4. www.itu.int/rec/T-REC-G.698.2/en
5. http://openroadm.org/home.html
6. T. Mizuochi, T. Sugihara, Y. Miyata, K. Kubo, K. Onohara, S. Hirano, H. Yoshida, T. Yoshida, and T. Ichikawa, "Evolution and status of forward error correction," Optical Fiber Communication, p. 6, 2012, OTu2A.
7. A. Leven and L. Schmalen, "Status and recent advances on forward error correction technologies for lightwave systems," *Journal of Lightwave Technology*, vol. 32, no. 16, pp. 2735–2750, 2014.
8. S. ten Brink and A. Leven, "FEC and soft decision: Concept and directions," Optical Fiber Communication, OW1H. p. 5, 2012.

10 Concluding Remarks

A diverse set of topics related to the introduction of coherent optics in the access environment has been reviewed. Given coherent optics technology development, cost-reduction trends, capacity demand trends and existing fiber-network assets of telcos, and cellular and cable companies, it is a natural evolution for coherent optics technology to migrate to the access environment. This book reviews the current state of access networks, the state of fiber deployment and potential new deployment strategies that can serve as a platform to introduce coherent optics technology in the access.

Challenges in optimizing the use of analog optics and balancing trade-offs of high carrier-to-noise ratio performance requirements at reasonable optical power levels to multiplex many wavelengths have led to cable's transition to distributed access architectures and the use of digital direct-detection systems. Different digital direct-detection systems such as high-level pulse amplitude modulation (PAM), Carrier-less Amplitude Phase (CAP) and Discrete Multitone (DMT) systems were discussed, including systems leveraging Kramers–Krönig receivers that trades in bandwidth to lower complexity and enhance receiver's sensitivity.

The increasing demand for bandwidth, the technical and cost trends of coherent optics, and the access environment shorter links provide the opportunity to redesign, optimize and simplify the implementations of coherent optics in the access. As a part of this redesign process, this book assesses changes in digital signal processing (DSP) functions, impact of soft decision forward error correction (SD FEC), power-consumption implications and the opportunity to leverage the ample link margin for the improvement of performance and the relaxation of component requirements.

The role of point-to-point (P2P) coherent optical links in access networks is examined, initially as a backhaul tool for broadband node, enterprise and base-station traffic and ultimately to provide direct-connectivity to high-capacity end-points. The transition of direct-detection passive optical networks (PONs) to coherent-based PONs including simplified coherent PON implementation as a path to bring the benefits of coherent technology to residential customers is explored.

In the quest to reduce costs, reduction of power consumption while increasing or maintaining performance and use of transceiver photonic-integrated systems technologies are necessary. The implementations of photonic-integrated systems on Silicon Photonics (SiPh) are compared to and contrasted with implementations on Indium-Phosphide (InP). Complementary metal–oxide–semiconductor (CMOS) compatibility, capability of integrating laser sources and optical amplifiers, size of components and loss through waveguides are some of the many elements that determine optimal selection of technology.

The different organizations developing specifications that address link distances below 120 km highlight the interest in coherent optics solutions from diverse industry segments, including data center, telco, cable and multi-industry perspectives.

A key differentiator among some specifications is the FEC selection which not only optimizes performance but also defines interoperability. An important question is whether the different specifications have enough scale to warrant industry-specific silicon implementations. While the economies of scale are promising, at an early phase in particular, a common silicon implementation is expected to address multiple standards. Favorable coherent component cost-reduction trends are expected to continue, technological advancements will enable higher performance and simpler implementations will make coherent technology more pervasive in the access so that exponential growth in capacity is met. The question is not whether coherent optics will migrate to the access but when will coherent optics technology be available in the access networks. Given the headway achieved in specification generation bodies and development progress of optical component and transceiver manufacturers focusing on shorter link distances, a future with coherent optics in the access is upon us.

Abbreviations

3GPP:	3rd generation partnership project
5G:	5th generation of cellular network technology
ADC:	analog-to-digital converter
ADSL:	asymmetric digital subscriber line
AES:	advanced encryption standard
AM:	amplitude modulation
APC:	angle-polished connector
APD:	avalanche photodiode
AR:	augmented reality
ASE:	amplified spontaneous emission
ASIC:	application-specific integrated circuit
AWG:	arbitrary waveform generator
AWGN:	additive white Gaussian noise
BER:	bit error rate
BPD:	balanced photodiode
bps:	bits per second
BPSK:	Binary phase-shift keying
CAP:	carrier-less amplitude/phase
CAUI:	100G attachment unit interface; (C = 100 Roman numeral)
CBR:	constant-bit-rate
CCAP:	converged cable access platform
CD:	chromatic dispersion
CDC:	chromatic dispersion compensation
cFEC:	concatenated forward error correction
CFO:	carrier frequency offset
CFP:	C-form factor pluggable
CMA:	constant modulus algorithm
CMMA:	cascaded multi-modulus algorithm
CMOS:	complementary metal-oxide-semiconductor
CMTS:	cable modem termination system
CNR:	carrier-to-noise ratio
CO:	central office
CP:	cyclic prefix
CRC:	cyclic redundancy check
CWDM:	coarse wavelength division multiplexing
DAA:	distributed access architecture
DAC:	digital-to-analog converter
dB:	decibel
dBm:	dB milliwatt
DC:	direct current
DCF:	dispersion compensation fiber

DCI:	data center interconnection
DCM:	dispersion compensation module
DCO:	digital coherent optics
DD:	direct detection
DDM:	direction-division multiplexing
Demux:	demultiplexer
DFB:	distributed feedback (laser)
DFE:	decision feedback equalizer
DFT:	discrete Fourier transform
DFTS:	discrete-Fourier-transform-spread
DL:	downlink
DMF:	dispersion managed fiber
DML:	directly modulated laser
DMT:	discrete multi-tone
DOCSIS:	data over cable service interface specification
DP-QPSK:	dual-polarization quadrature phase-shift keying
DSB:	double sideband
DSL:	digital subscriber line
DSLAM:	digital subscriber line access multiplexer
DSP:	digital signal processing
DVB-T:	digital video broadcasting-terrestrial
DWDM:	dense wavelength division multiplexing
E2E:	end-to-end
EAM:	electro-absorption modulator
ECC:	elliptic-curve cryptography
ECL:	external cavity laser
EDFA:	erbium-doped fiber amplifier
EML:	electro-absorption modulated laser
ENOB:	effective number of bits
EPON:	Ethernet passive optical network
ESS:	Ethernet security specifications
ETDM:	electrical time division multiplexing
EVM:	error vector magnitude
FBMC:	filter-bank multi-carrier
FEC:	forward error correction
FFE:	feed-forward equalizer
FFT:	fast-Fourier transform
FIPS:	Federal Information Processing Standard
FIR:	finite impulse response
FPGA:	field-programmable gate array
FWM:	four-wave mixing
FTTdp:	fiber to the distribution point
FTTH:	fiber to the home
Gbps:	gigabit per second
GCM:	Galois counter mode
Ge:	germanium

GFDM:	generalized frequency-division multiplexing
GHz:	gigahertz
GMAC:	Galois message authentication code
HD:	high definition
HDSL:	high-speed digital subscriber line
HFC:	hybrid fiber-coax
HHP:	household pass
Hz:	hertz
I:	in-phase
IA:	implementation agreement
ICI:	inter-channel interference
ICV:	integrity check value
IEEE:	Institute of Electrical and Electronics Engineers
IFFT:	inverse fast-Fourier transform
IM-DD:	intensity-modulation and direct-detection
InP:	indium-phosphide
IoT:	Internet of Things
IP:	Internet protocol
IPSec:	IP security
I/Q:	in-phase and quadrature
ISBE:	International Society of Broadband Experts
ISI:	inter-symbol interference
ITU:	International Telecommunication Union
IV:	initialization vector
km:	kilometer
LAN:	local area network
LCP:	local convergence point
LD:	laser diode
$LiNbO_3$/LN:	lithium-niobate
LMS:	least mean square
LO:	local oscillator
LPF:	low-pass filter
LTE:	long-term evolution
LUT:	look up table
MAC:	media access control
MACSec:	media access control security
MAN:	metro area network
MHz:	megahertz
MIMO:	multi-input–multi-output
MKA:	media access control security key agreement
MLSE:	maximum likelihood sequence estimation
MMI:	multi-mode interference
MPI:	multiple-path interference
MPLS:	multiprotocol label switching
MSA:	multi-source agreement
MSO:	multi-system operator

Mux: multiplexer
MZI: Mach–Zehnder interferometer
MZM: Mach–Zehnder modulator
NCG: net coding gain
NIR: near-infrared
NL: nonlinear
NRZ: non-return zero
NSA: National Security Agency
NZDSF: non-zero dispersion shifted fiber
OFDM: orthogonal frequency division multiplexing
OIF: Optical Internetworking Forum
OLT: optical line terminal
ONU: optical network unit
OOB: out of band
OOK: on–off keying
OPLL: optical phase-locked loop
OSFP: Octal Small Formfactor Pluggable
OSNR: optical signal-to-noise ratio
OTN: optical transport network
P2MP: point-to-multipoint
P2P: point-to-point
PAM: pulse amplitude modulation
PARP: peak-to-average-power ratio
PBC: polarization beam combiner
PBS: polarization beam splitter
PDL: polarization-dependent loss
PHY: physical layer
PIC: photonic integration circuit
PIN/p–i–n: positive–intrinsic–negative
PKI: public key infrastructure
PM: polarization multiplexing
PMD: polarization mode dispersion
PON: passive optical network
POTS: plain old telephone service
P-OTS: packet-optical transport systems
PPG: pulse pattern generator
PR: phase recovery
PSK: phase-shift keying
Q: in-quadrature
Q: quality factor
QAM: quadrature amplitude modulation
QPSK: quadrature phase-shift keying
eQSFP: quad small form-factor pluggable
QSFP-DD: quad small form-factor pluggable double density
R-MFL: receiver matched filter length
R-PHY: remote PHY

RF: radio frequency
RFoG: Radio Frequency over glass
RIN: relative intensity noise
ROADM: open reconfigurable optical add/drop multiplexers
RPD: remote PHY device
RS: Reed–Solomon
Rx: receiver
SBS: stimulated Brillouin scattering
SCM: sub-carrier modulation
SCTE: Society of Cable Telecommunications Engineers
SD-FEC: soft-decision forward error correction
SE: spectral efficiency
Ser/Des: serializers/deserializers
SG-DBR: sampled grating distributed Bragg reflector
SiPh: silicon photonics
SMF: single-mode fiber
SNR: signal-to-noise ratio
SOA: semiconductor optical amplifier
SOI: silicon on insulator
SOP: state of polarization
SPM: self-phase modulation
SSB: single sideband
SSBI: signal-to-signal-beat interference
TEC: thermo-electric cooler
TFF: thin-film filter
TIA: transimpedance amplifier
T-OFL: transmitter orthogonal filter length
UFMC: universal-filtered multi-carrier
UL: uplink
ULA: ultra-large area
ULLF: ultra-low loss
VCSEL: vertical cavity surface-emitting laser
VDSL: very high speed DSL
VLAN: virtual local area network
VR: virtual reality
VSB: vestigial sideband
WAN: wide area network
WDM: wavelength-division multiplexing
WiMAX: worldwide interoperability for microwave access
WLAN: wireless local access network
WPAN: wireless personal area network
XPM: cross-phase modulation

Index

Printed and bound by CPI Group (UK) Ltd, Croydon, CR0 4YY

17/10/2024

01775682-0019